Math for the World of Work

Student Workbook

by
Kathleen Harmeyer

American Guidance Service, Inc.
Circle Pines, Minnesota 55014-1796
1-800-328-2560

© 2002 AGS® American Guidance Service, Inc. Circle Pines, MN 55014-1796.
All rights reserved, including translation. No part of this publication may be reproduced or transmitted in any form or by any means without written permission from the publisher.

Printed in the United States of America

ISBN 0-7854-2700-7

Product Number 93403

A 0 9 8 7 6 5 4 3 2 1

Nichi

Table of Contents

Workbook 1 Wages
Workbook 2 Time Worked
Workbook 3 Other Kinds of Payment
Workbook 4 Annual Income
Workbook 5 Commissions
Workbook 6 More Commissions Practice
Workbook 7 Health Benefits
Workbook 8 Insurance Benefits
Workbook 9 Retirement Benefits
Workbook 10 More Retirement Benefits Practice
Workbook 11 Profit Sharing, Employee Ownership, Stock Options
Workbook 12 Manufacturing
Workbook 13 More Manufacturing Practice
Workbook 14 Sales
Workbook 15 Retail
Workbook 16 Service
Workbook 17 Nonprofit Businesses
Workbook 18 Staffing
Workbook 19 Hiring
Workbook 20 Employee Performance
Workbook 21 Labor Issues
Workbook 22 More Labor Issues Practice
Workbook 23 Comparing Airfares
Workbook 24 More Comparing Airfares Practice
Workbook 25 Selecting Lodging
Workbook 26 Using Credit Cards
Workbook 27 Travel Reimbursement Forms
Workbook 28 Simple Interest
Workbook 29 Compound Interest
Workbook 30 Business Loans

Workbook 31 Cash Flow
Workbook 32 Product Payments
Workbook 33 More Product Payments Practice
Workbook 34 Salaries and Benefits
Workbook 35 Building Facilities
Workbook 36 Transportation
Workbook 37 Cost of Production
Workbook 38 More Cost of Production Practice
Workbook 39 Cost of Sales
Workbook 40 Inventory
Workbook 41 Profit and Loss
Workbook 42 Kinds of Insurance
Workbook 43 Cost of Insurance
Workbook 44 More Cost of Insurance Practice
Workbook 45 Insurance as an Investment
Workbook 46 Environmental Regulations
Workbook 47 Americans with Disabilities Act
Workbook 48 Interstate Commerce
Workbook 49 International Business
Workbook 50 Financial Risks
Workbook 51 Legal Risks
Workbook 52 Sales and Revenue Projections
Workbook 53 Conventions and Exhibits
Workbook 54 More Conventions and Exhibits Practice
Workbook 55 Marketing Channels
Workbook 56 Catalogs and Customer Service
Workbook 57 Shipping Orders
Workbook 58 Handling and Processing Orders
Workbook 59 Inventory and Warehousing
Workbook 60 More Inventory and Warehousing Practice

Name Date Period

Chapter 1, Lesson 1

Workbook

1

Wages

EXAMPLE Joan earns \$9.20 per hour. She works $18\frac{1}{2}$ hours per week. What are her weekly wages?

hourly rate × hours worked = wages
\$9.20 per hour × $18\frac{1}{2}$ hours = wages

Rewrite $\frac{1}{2}$ as a decimal.
$\frac{1}{2} = 1 \div 2 = .5$ $18\frac{1}{2} = 18.5$

$\$9.20 \times 18.5 = \$170.20$

Joan's weekly wages are \$170.20.

Directions: Find the weekly wages.

	Hourly Rate	Hours Worked	Weekly Wages
1.	\$6.80	35	
2.	\$10.50	16	
3.	\$12.45	25	
4.	\$9.00	$24\frac{3}{4}$	
5.	\$15.00	$30\frac{1}{2}$	

Directions: Estimate the annual wages. Use 50 to estimate the number of weeks in a year.

	Hourly Rate	Hours Worked Each Week	Estimated Annual Wages
6.	\$8.70	29	
7.	\$7.10	41	
8.	\$9.25	21	
9.	\$12.30	17	
10.	\$19.76	26	

Directions: Find the net pay.

	Gross Pay	DEDUCTIONS			Net Pay
		Taxes	Retirement Account	Insurance	
11.	\$347.00	\$27.00	\$18.00	\$4.00	
12.	\$218.50	\$21.50	\$10.00	\$3.50	
13.	\$391.00	\$36.83	\$22.50	\$6.25	
14.	\$450.85	\$41.50	\$27.45	\$7.75	
15.	\$243.19	\$19.62	\$12.35	\$12.40	

Timed Worked

EXAMPLE Ernie starts work at 6:25 A.M. He goes to lunch at 11:15 A.M. and returns at 1:00 P.M. He leaves work at 4:50 P.M. How many hours does Ernie work?

Step 1 Subtract to find the number of hours worked. Rename 1 hour as 60 minutes. Add 60 minutes to 15 minutes.

~~11:15~~ 10:75 A.M. (later time)	4:50 P.M. (later time)
− 6:25 A.M. (earlier time)	− 1:00 P.M. (earlier time)
4:50 = 4 hr., 50 min.	3:50 = 3 hr., 50 min.

Step 2 Find the total hours worked.

4 hours, 50 minutes
+ 3 hours, 50 minutes
7 hours, 100 minutes

Step 3 Rename 7 hours, 100 minutes. Write the minutes as a fraction of an hour.

100 min. = 60 min. + 40 min. = 1 hr., 40 min.
7 hr., 100 min. = 8 hr., 40 min.

$\frac{40 \text{ min.}}{60 \text{ min.}}$ Minutes worked / Minutes in 1 hour $\frac{40 \div 20}{60 \div 20} = \frac{2}{3}$

Ernie works $8\frac{2}{3}$ hours.

Directions: Find the number of hours worked each day.

	Day	In	Out	In	Out	Daily Hours
1.	Mon.	6:00 A.M.	11:00 A.M.	1:00 P.M.	4:00 P.M.	
2.	Tue.	6:00 A.M.	11:45 A.M.	1:00 P.M.	4:30 P.M.	
3.	Wed.	9:00 A.M.	11:30 A.M.	1:00 P.M.	4:15 P.M.	
4.	Thurs.	8:15 A.M.	12:00 P.M.	1:00 P.M.	3:45 P.M.	
5.	Fri.	7:30 A.M.	11:45 A.M.	1:00 P.M.	5:30 P.M.	

Directions: Calculate yearly income. Use time and a half for the overtime rate when the work is more than 40 hours per week.

	Hourly Rate	Hours per Week	Weeks Worked	Yearly Income
6.	$5.40	30	40	
7.	$10.00	45	50	
8.	$6.40	42	48	
9.	$12.50	48	35	
10.	$16.80	44	49	

Name Date Period

Other Kinds of Payment

EXAMPLE Maria parks cars for a valet service. She earns $8.25 per hour. One week she worked 26 hours and received $119.00 in tips. What was her total income for the week?

$ 8.25	Hourly rate	$ 214.50	Hourly wages
× 26	Hours worked	+119.00	Tips
$214.50	Hourly wages	$ 333.50	Total income

Maria's total income for the week was $333.50.

Directions: Find the total income.

	Hours Worked	Hourly Rate	Tips Received	Total Income
1.	36	$7.50	$123.00	
2.	27	$5.95	$87.00	
3.	18	$9.30	$162.25	
4.	45	$8.10	$138.50	

Directions: Find the weekly totals and the wages for each employee.

	Employee	Daily Production					Total	Weekly Rate	Piecework Wages
		M	Tu	W	Th	F			
5.	Ted	31	29	30	26	35		$3.06	
6.	Alicia	5	6	8	9	4		$11.45	
7.	Larry	12	8	15	11	10		$6.41	
8.	Oscar	41	32	35	28	36		$2.88	

Directions: Solve.

9. Sonia clears and sets tables. She earns $6.80 per hour plus 10% of the waiters' tips. If the waiters make $274 in tips during her 8-hour shift, how much does Sonia earn? ________

10. Jason assembles oxygen masks. He earns $11.45 per hour plus $.03 per mask. One day he worked 7 hours and made 2,368 masks. How much did Jason earn? ________

Name Date Period

Annual Income

EXAMPLE Denise earns $9.15 per hour doing data entry. If she works 32 hours a week, what is her annual income?

Step 1 Find the number of hours worked in a year.

52	Weeks worked per year
× 32	Hours worked per week
1,664	Hours worked in a year

Step 2 Multiply to find the annual income.

$ 9.15	Hourly rate
× 1,664	Hours worked in a year
$15,225.60	Annual income

Denise's annual income is $15,225.60.

Directions: Each person earns $9.15 per hour. Find each annual income.

	Employee	Hours Worked per Week	Annual Income
1.	Gina	15	
2.	Susan	21	
3.	Daniel	30	
4.	Niki	24	
5.	Steve	19	

Directions: Find the number of pay periods per year for each employee.

	Employee	How Often Paid	Pay Periods per Year
6.	Tom	Monthly	
7.	Irene	Bimonthly	

Directions: Find the number of pay periods per year and the earnings per pay period. Round to the nearest cent.

	Annual Salary	How Often Paid	Pay Periods per Year	Earnings per Pay Period
8.	$72,582.60	Weekly		
9.	$42,350.00	Biweekly		
10.	$58,645.50	Quarterly		

Name Date Period

Commissions

EXAMPLE Thad sells $480,000 in stocks to clients. His rate of commission is 1.8%. What is Thad's commission?

Commission = Sales × Rate of commission
= $480,000 × 1.8%

Change the percent to a decimal. Multiply.
$480,000 × .018 = $8,640.00

Thad's commission for selling stocks is $8,640.

Directions: Find the amount of commission for each amount of sales.

	Amount of Sales	Rate of Commission	Amount of Commission
1.	$20,000	7%	
2.	$60,700	9%	
3.	$230,000	4.5%	
4.	$350,000	10%	
5.	$186,500	6.5%	
6.	$45,100	5%	
7.	$92,300	8.2%	
8.	$271,200	3.9%	

Directions: Find the total sales needed to reach each income goal. Round to the nearest dollar.

	Income Goal	Rate of Commission	Total Sales Needed to Reach Goal
9.	$6,000	4%	
10.	$12,000	12%	
11.	$5,200	7%	
12.	$19,381	9%	
13.	$23,000	8%	
14.	$26,500	4.5%	
15.	$18,000	6.5%	

Name Date Period

More Commissions Practice

EXAMPLE Barbara sells ads for a monthly salary of $1,350. She also earns a commission of 3% on her total sales. Last month she sold $184,000 in ads. What was her income for last month?

$ 184,000	Total sales	$1,350	Monthly salary
× .03	Rate of commission	+5,520	Commission
$5,520.00	Commission	$6,870	Total income

Barbara's income was $6,870 last month.

Directions: Find commission and total earnings.

	Total Sales	Rate of Commission	Salary Earned	Commission	Total Earnings
1.	$40,000	8%	$1,150		
2.	$60,000	5%	$850		
3.	$24,000	10%	$2,100		
4.	$85,000	3%	$900		
5.	$64,000	6%	$1,320		
6.	$150,000	2%	$1,100		
7.	$228,500	1.5%	$2,300		
8.	$161,000	8.25%	$650		

Directions: Find the total income.

	Salary	Bonus	Total Sales	Total Income
9.	$47,000	5% of sales over $100,000	$138,000	
10.	$35,800	12% of sales over $100,000	$185,000	
11.	$29,000	50% of sales over $150,000	$215,000	
12.	$51,500	25% of sales over $60,000	$93,000	
13.	$63,000	15% of sales over $92,000	$171,000	

Directions: Solve.

14. Which is more, 9% of $2,000 or 12% of $850? How much more? ________

15. Jason earns a 4% commission on sales of $6,000. Pete earns a 2% commission on sales of $11,000. Who earns more in commission? How much more? ________

Name Date Period

Health Benefits

EXAMPLE Iris's health insurance policy pays 80% of her medical bills, and she pays the remaining 20%. Last year she had $7,100 in doctor bills and $10,055 in hospital bills, and she paid $2,340 for medicines. What is Iris's co-payment? What is the benefit paid by the insurance company?

Step 1 Find Iris's total medical expenses. $7,100 + $10,055 + $2,340 = $19,495

Step 2 Find 20% of Iris's total medical expenses. $19,495 × .2 = $3,899

Step 3 Subtract the co-payment from the total medical expenses. $19,495 − $3,899 = $15,596

Iris's co-payment is $3,899, and the benefit is $15,596.

Directions: Each co-payment is 20% of costs. Find the co-payment and benefit amounts.

	Hospital Bills	Doctors' Bills	Medicines	Co-payment	Benefit
1.	$11,200.00	$0	$115.00		
2.	$4,200.00	$1,125.00	$298.00		
3.	$31,400.00	$0	$3,072.00		
4.	$19,508.50	$8,361.50	$2,006.70		
5.	$41,307.80	$0	$3,744.20		
6.	$0	$3,008.60	$92.50		
7.	$48,572.30	$0	$7,374.15		
8.	$19,651.45	$6,700.55	$1,892.00		
9.	$0	$2,992.30	$61.40		
10.	$72,398.20	$21,685.80	$8,600.00		
11.	$123,200.60	$37,946.50	$11,257.30		
12.	$91,274.10	$42,883.20	$13,080.20		

Directions: Solve problems 13–15.

13. Martha has a $200 deductible and 20% co-payment on her health insurance. Last year she had $6,500 in medical bills. What was Martha's out-of-pocket expense? __________

14. Mitch has an $800 deductible and no co-payment on his health insurance. If he had a total of $4,092 in doctor's bills, what is the amount of his benefit? __________

15. If you have medical expenses of $7,320 in 1 year, which costs less, a $600 deductible and 10% co-payment, or an $800 deductible and 5% co-payment? __________

Name Date Period

Insurance Benefits

EXAMPLE Rachel earns $90,000 a year at Center City Corporation. If she dies at age 68, how much money will her beneficiary receive from her life insurance policy?

Because Rachel is between 68 and 72 years old, her beneficiary will receive 45% of her annual salary.

$90,000 × .45 = $40,500

Rachel's beneficiary will receive $40,500 upon her death.

Employee's Age	Coverage Level
55 or under	100%
56–62	80%
63–67	55%
68–72	45%
73 or over	35%

Directions: Find the life insurance benefit for each employee. Use the coverage levels in the table above.

	Employee	Age at Death	Annual Salary	Benefit
1.	S. Garcia	74	$92,000	
2.	D. Cline	58	$71,000	
3.	M. Williams	71	$48,000	
4.	T. Young	70	$61,000	
5.	L. Smith	61	$86,000	
6.	W. Thatcher	59	$95,000	

Directions: Use the life expectancy table at the right to solve problems 7–10.

7. How much greater is the life expectancy of a male born in 1993 than that of a male born in 1985?

____

8. How much greater is the life expectancy of a female born in 1990 than that of a female born in 1982?

____

9. If you are a female 18 years, 3 months old and were born in 1989, how much longer are you expected to live?

____

10. What is the difference between male and female life expectancy for those born in 1987?

____

Year of Birth	Total	Male	Female
1982	74.5	70.9	78.1
1983	74.6	71.0	78.1
1984	74.7	71.2	78.2
1985	74.7	71.2	78.2
1986	74.8	71.3	78.3
1987	75.0	71.5	78.4
1988	74.9	71.5	78.3
1989	75.1	71.7	78.5
1990	75.3	71.8	78.8
1991	75.5	72.0	78.9
1992	75.5	72.1	78.9
1993	75.5	72.2	78.8
1994	75.7	72.4	79.0

Name Date Period

Retirement Benefits

EXAMPLE Kuong paid $60,000 in Social Security taxes over 30 years. His monthly Social Security benefit is about $1\frac{1}{2}$% of the total contributions. Estimate his monthly Social Security benefit.

Kuong's contribution + Employer's contribution = Total

↓ ↓ ↓

$60,000 + $60,000 = $120,000

$\$120{,}000 \times 0.15 = \$1{,}800$

If Kuong retires at age 65, his monthly Social Security benefit is about $1,800.

Directions: Use the formula $B \approx .015C$ to estimate each monthly benefit. Round to the nearest dollar.

	Employee Contribution	Employer Contribution	Total Contribution	Retirement Benefit (based on $1\frac{1}{2}$%)
1.	$40,000			
2.	$50,000			
3.	$90,000			
4.	$75,000			
5.	$28,500			
6.	$38,920			
7.	$61,308			
8.	$57,091			

Directions: Complete. Use $1\frac{1}{2}$% of total Social Security contributions to calculate monthly benefit. Round to the nearest dollar.

	Total Contribution	Monthly Benefit	Annual Benefit	Years of Retirement	Total Retirement Benefit	Benefits Greater or Less Than Contributions?
9.	$42,000			19		
10.	$50,000			6		
11.	$91,050			2		
12.	$57,100			7		
13.	$46,230			5		
14.	$73,258			10		
15.	$81,191			$3\frac{1}{2}$		

More Retirement Benefits Practice

EXAMPLE Carla worked for the same company for 20 years and retired at age 65. Her average monthly pay was $2,800. Her Social Security benefit is $800 a month. Calculate Carla's monthly pension, using the following steps.

Step 1 Find 60% of average monthly pay. $.6\% \times \$2,800 = \$1,680$

Step 2 Find 50% of monthly Social Security benefit. $.5\% \times \$800 = \$400$

Step 3 Subtract Step 2 from Step 1. $\$1,680 - \$400 = \$1,280$

Step 4 Divide years worked for the company (up to 35) by 35. $20 \div 35 \approx .571$

Step 5 Multiply Step 3 by Step 4. $\$1,280 \times .571 = \$730.88 \approx \$731$

Carla's monthly pension from her company is $731.

Directions: Use steps 1–5 above to find each monthly pension benefit. Round decimals to the nearest thousandth. Round money to the nearest dollar.

	Average Monthly Pay	Social Security Benefit	Years Worked	Years Worked Divided by 35	Monthly Pension Benefit
1.	$3,000	$1,600	17		
2.	$5,000	$2,300	25		
3.	$2,400	$1,000	30		
4.	$3,600	$1,640	24		
5.	$3,850	$1,850	12		
6.	$4,700	$2,100	15		
7.	$2,925	$1,536	34		
8.	$3,275	$1,700	29		

Directions: Complete to find each total annual 401(k) contribution.

	Earnings Deferred per Paycheck	Pay Schedule	Annual Employee Contribution	Employer Matching Plan	Employer Contribution	Total Annual Contribution
9.	$250	Weekly		1:1		
10.	$100	Biweekly		2:1		
11.	$209	Weekly		2:1		
12.	$350	Semimonthly		4:1		
13.	$641	Monthly		2:1		
14.	$482	Bimonthly		4:1		
15.	$92.50	Weekly		2:1		

Name Date Period

Profit Sharing, Employee Ownership, Stock Options

EXAMPLE A company has a profit-sharing plan that distributes 10% of its annual profits equally among 200 employees. If the company makes a profit of $2,600,000, how much does each employee receive in profit-sharing benefits?

Profit × Percent shared = Profit shared with employees

$2,600,000 × .10 = $260,000

Profit shared ÷ Number of employees = Profit each employee receives

$260,000 ÷ 200 = $1,300

Each employee receives $1,300 of the company's profit.

Directions: Find the profit-sharing benefit per employee. Round to the nearest dollar.

	Annual Profit	Percent of Profit Shared	Number of Employees	Benefit per Employee
1.	$940,000	20%	40	
2.	$1,800,000	10%	100	
3.	$620,000	15%	225	
4.	$5,400,000	20%	347	
5.	$3,610,000	15%	92	

Directions: Estimate, then find the actual profit.

	Purchase Price	Selling Price	Estimated Profit	Amount of Profit
6.	$.72	$2.13		
7.	$2.61	$9.82		
8.	$12.20	$21.36		
9.	$28.71	$46.80		
10.	$43.90	$102.09		

Directions: Find the profit from exercising the given number of options.

	Purchase Price	Selling Price	Number of Options Exercised	Profit
11.	$4.00	$12.00	400	
12.	$19.00	$30.00	380	
13.	$13.29	$27.64	2,500	
14.	$29.61	$41.02	900	
15.	$38.75	$52.29	1,300	

Name Date Period

Chapter 3, Lesson 1

Workbook 12

Manufacturing

EXAMPLE A company receives an order for 112 clocks. It can make 7 clocks per day. Shipping takes 2 weeks. The customer wants the order by October 10. When should production begin?

7 clocks per day × 5 work days per week = 35 clocks per week

112 clocks ordered ÷ 35 clocks per week = 3.2 weeks

.2 of a week × 5 work days = 1 day

3 weeks, 1 work day + 2 weeks for shipping = 5 weeks, 1 work day

Count back 5 weeks, 1 work day from October 10. This date is September 4, which is a holiday. Count back another work day, to September 1.

Production should begin on September 1.

Directions: Find production time. Then use the calendar below to find the date each production period should begin. Allow 2 weeks' shipping time on all orders.

SEPTEMBER

S	M	T	W	T	F	S
					1	2
3	4	5	6	7	8	9
10	11	12	13	14	15	16
17	18	19	20	21	22	23
24	25	26	27	28	29	30

OCTOBER

S	M	T	W	T	F	S
1	2	3	4	5	6	7
8	9	10	11	12	13	14
15	16	17	18	19	20	21
22	23	24	25	26	27	28
29	30	31				

NOVEMBER

S	M	T	W	T	F	S
			1	2	3	4
5	6	7	8	9	10	11
12	13	14	15	16	17	18
19	20	21	22	23	24	25
26	27	28	29	30		

Dates in gray are holidays.

	Number Ordered	Number Produced per Day	Number Produced per Week	Production Time (weeks and days)	Delivery Date	Production Start Date
1.	30 grandfather clocks	2			Nov. 15	
2.	1,200 alarm clocks	40			Nov. 14	
3.	600 men's watches	120			Nov. 8	
4.	600 women's watches	100			Sept. 29	
5.	675 analog clocks	75			Nov. 8	
6.	1,012 digital clocks	92			Oct. 6	
7.	1,904 clock faces	238			Nov. 20	
8.	55 clock billboards	5			Oct. 4	
9.	5,015 travel clocks	295			Nov. 10	
10.	18 schoolhouse clocks	3			Sept. 26	

Name Date Period

More Manufacturing Practice

EXAMPLE Manufacturing one skateboard requires 1 deck (board), 2 trucks (axles), 4 wheels, 8 bearings, 4 bolts, and 1 grip. How many of each are needed to make 20 skateboards?

Multiply the amount needed for 1 skateboard by 20.

1 deck per skateboard × 20 skateboards = 20 decks
2 trucks per skateboard × 20 skateboards = 40 trucks
4 wheels per skateboard × 20 skateboards = 80 wheels
8 bearings per skateboard × 20 skateboards = 160 bearings
4 bolts per skateboard × 20 skateboards = 80 bolts
1 grip per skateboard × 20 skateboards = 20 grips

Directions: Find the materials needed to manufacture each given number of skateboards.

	Number of Skateboards	Decks Needed	Trucks Needed	Wheels Needed	Bearings Needed	Bolts Needed	Grips Needed
	1	1	2	4	8	4	1
1.	86						
2.	723						
3.	2,400						
4.	3,580						
5.	6,058						
6.	10,251						

Directions: Solve problems 7–10 using 1 deck, 2 trucks, 4 wheels, 8 bearings, 4 bolts, and 1 grip for each skateboard.

7. There are 920 wheels in stock. How many more wheels are needed to complete production of 2,200 skateboards? ______

8. There are 6,200 bearings in stock. How many skateboards can be made with 6,200 bearings? ______

9. There are 418 trucks in stock. How many more trucks are needed to make 2,852 skateboards? ______

10. There are 754 decks, 2,980 trucks, 4,200 wheels, 7,000 bearings, 3,500 bolts, and 950 grips in stock. How many more items of each type are needed to fill an order for 921 skateboards? ______

Sales

EXAMPLE A company sets a goal of $840,000 in annual revenue from fax machines. Each fax machine sells for $600. What is the average number of fax machines that must be sold monthly to reach the annual revenue goal?

Step 1 Divide the annual revenue goal by the selling price of 1 fax machine.

$840,000 ÷ $600 = 1,400

Step 2 Divide the annual sales by 12. Round up.

1,400 ÷ 12 ≈ 117

The company must sell an average of 117 fax machines monthly.

Directions: Find the number of sales needed to meet each revenue goal.

	Annual Revenue Goal	Income from Each Sale	Annual Sales Goal	Monthly Sales Goal
1.	$336,000	$700		
2.	$2,484,000	$900		
3.	$250,560	$348		
4.	$2,723,105	$847		
5.	$162,032	$328		

Directions: Find each discounted price and the profit lost.

	Regular Price	Discount Percent	Discount Price	Profit Lost
6.	$60.00	20%		
7.	$250.00	30%		
8.	$24.50	10%		
9.	$62.90	40%		
10.	$485.00	25%		

Directions: Find the total profit lost.

	Regular Price	Discount Percent	Number of Discounted Items	Profit Lost
11.	$90.00	30%	500	
12.	$400.00	20%	200	
13.	$230.00	60%	340	
14.	$92.10	40%	120	
15.	$48.20	15%	92	

Retail

EXAMPLE The Gas Express schedules 1 cashier for every 120 customers. The table shows the average number of customers at different times. How many cashiers are scheduled to work Sundays from 9 P.M. to 3 A.M.?

Shift	Hours	Average Number of Customers		
		Mon.–Fri.	Sat.	Sun.
Shift 1	9 A.M.– 3 P.M.	240	280	570
Shift 2	3 P.M.– 9 P.M.	412	351	492
Shift 3	9 P.M.– 3 A.M.	305	412	135
Shift 4	3 A.M.– 9 A.M.	296	86	261

Divide the number in the *Shift 3* row and *Sun.* column by 120. Round up to have enough cashiers.

$135 \div 120 = 1.125 \rightarrow 2$

Gas Express schedules 2 cashiers to work between 9 P.M. and 3 A.M. on Sundays.

Directions: Use the table above to find the number of cashiers needed.

Shift	Mon.–Fri.	Sat.	Sun.
Shift 1	**1.**	**2.**	**3.**
Shift 2	**4.**	**5.**	**6.**
Shift 3	**7.**	**8.**	2
Shift 4	**9.**	**10.**	**11.**

Directions: An appliance store buys stoves for $319 each and sells them for $2,120 each. Solve problems 12–15.

12. The store discounts the price 30%. What is the discounted price? ________

13. The store usually sells 5 stoves a week at the regular price. When it discounts the price, sales increase by about 80%. How many stoves a week does it expect to sell at the discounted price? ________

14. What is the weekly profit from selling stoves at the regular price? What is the weekly profit from selling stoves at the discounted price? ________

15. In how many weeks could the store sell 92 stoves at a 30% discount? How much profit would the store make? ________

Name Date Period

Service

EXAMPLE Chu's Dusters uses about 8 ounces of cleaner to clean 1 house. Chu's cleans 379 houses a week. About how many 32-ounce bottles of cleaner are needed for one week's work?

Step 1 Find the number of houses that can be cleaned with a 32-ounce bottle.

32-ounce bottle ÷ 8 ounces per house = 4 houses

Step 2 Divide the number of houses cleaned weekly by the number cleaned using 1 bottle.

379 houses per week ÷ 4 houses per bottle = 94.75 bottles per week $94.75 \approx 95$

About 95 bottles of cleaner are needed for one week's work.

Directions: Find the supplies used each month. Round to the nearest whole number.

	Item	Average Amount Used	Houses Cleaned per Week	Amount Used per Week
1.	128-oz. bottle of cleaner	16 oz. per house	56	
2.	Box of 250 cleaning wipes	10 wipes per house	347	

Directions: Find the number of employees needed to clean each area.

	Size (feet by feet)	Area Cleaned by 1 Person in 1 Hour	Time Available (hours)	Number of Employees
3.	200 by 200	400 sq. ft.	5	
4.	40 by 120	200 sq. ft.	6	
5.	150 by 320	300 sq. ft.	8	
6.	60 by 250	750 sq. ft.	4	
7.	80 by 434	496 sq. ft.	7	
8.	120 by 500	800 sq. ft.	7.5	

Directions: Find how many boxes of trash can liners are used each week.

	Liners Used	Total Square Feet Cleaned Each Week	Liners in 1 Box	Boxes Used Each Week
9.	8 per 400 square feet	31,200	52	
10.	6 per 500 square feet	75,000	60	

Name Date Period

Nonprofit Businesses

EXAMPLE Food Pantry is a nonprofit business that gives food to needy people. The circle graph to the right shows 100% of its annual budget. In 2002 Food Pantry receives $400,000 from donations and fund-raising campaigns. How much money is budgeted in 2002 for canned goods?

Food Pantry Annual Budget

Multiply $400,000 by the percent budgeted for canned goods.

$\$400{,}000 \times .20 = \$80{,}000$

Food Pantry budgets $80,000 in 2002 for canned goods.

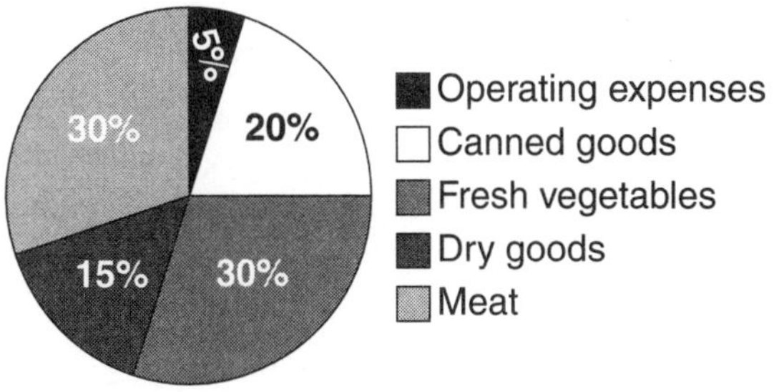

Directions: Use the Food Pantry Annual Budget. Find each budgeted amount.

	Income	Expense	Budgeted Amount
1.	$600,000	Fresh vegetables	
2.	$1,800,000	Dry goods	
3.	$460,000	Meat	
4.	$3,200,000	Operating expenses	

Directions: Use the Food Pantry Annual Budget to solve.

5. One year Food Pantry has an income of $500,000. How much money is budgeted for vegetables? ____________

Directions: Find the amount of money raised by each fund-raising campaign.

	Event	Cost	Number of Sales	Money Raised
6.	Pancake breakfast	$3 a ticket	2,152	
7.	Car wash	$15 per car	926	
8.	Candy sales	$2 per candy bar	8,709	
9.	Basketball game	$35 per ticket	4,563	

Directions: Use the fund-raising campaigns above to solve problem 10.

10. How much more did the candy sales raise than the car wash? ____________

Staffing

EXAMPLE Dress Manufacturers has a contract to supply 1,046,500 pairs of socks to a chain of stores over the next year. The plant is open 5 days a week. One worker can produce 35 pairs of socks per day. What production staff is needed to meet the production goal?

Step 1 Find how many socks need to be produced each day.
5 days per week × 52 weeks per year = 260 work days a year
1,046,500 pairs of socks ÷ 260 days = 4,025 pairs of socks per day

Step 2 Divide the daily production goal by 1 worker's daily production.
4,025 pairs of socks ÷ 35 pairs of socks = 115 workers

The manufacturer needs 115 workers.

Directions: Find the production staff needed to meet each annual goal in 260 work days. Round decimals up to meet goals.

	Kinds of Socks	Number Produced by 1 Worker in 1 Day	Annual Production Goal	Production Staff Needed
1.	Knit	50 pairs	780,000 pairs	
2.	Tube	40 pairs	946,400 pairs	
3.	Athletic	60 pairs	1,138,800 pairs	
4.	Booties	24 pairs	343,200 pairs	
5.	Children's	84 pairs	518,000 pairs	

Selected State Population Estimates			
State	**Population**	**State**	**Population**
Colorado	4,056,133	Oklahoma	3,358,044
Delaware	753,538	Texas	20,044,141
Louisiana	4,372,035	Wisconsin	5,250,446

Directions: Use the table above to find the sales force needed for each territory.

	State	Salesperson-to-Population Ratio	Sales Force Needed for Territory
6.	Wisconsin	1 to 15,000	
7.	Louisiana	1 to 9,000	
8.	Texas	1 to 50,000	
9.	Delaware	1 to 25,000	

Directions: Use the table above to solve problem 10.

10. A company wants 1 salesperson for every 50,000 people in the territory. It has a sales force of 70 people. For which state(s) can the company supply a large enough sales force? ______________

Hiring

EXAMPLE In 1999 the unemployment rate in North Dakota was 3.4%. By 2000, with a labor force of 337,000 people, the rate had fallen to 3.0%. How much did the unemployment rate change? How many people were unemployed in the year 2000? Round to the nearest thousand.

Step 1 Subtract the lesser rate from the greater rate.

$3.4\% - 3.0\% = .4\%$

The unemployment rate fell .4%, or changed $-.4\%$.

Step 2 Multiply the unemployment rate in 2000, written as a decimal, by the labor force in 2000. Round to the nearest thousand.

$337{,}000 \times .03 = 10{,}110$ $10{,}110 \approx 10{,}000$

About 10,000 people were unemployed in North Dakota in 2000.

Directions: Find the change in the unemployment rate between 1999 and 2000 for each state. Write + for an increase and − for a decrease.

	State	Labor Force, 2000	Unemployment Rate, 1999	Unemployment Rate, 2000	Change in Rate
1.	New York	8,889,000	5.2%	4.6%	
2.	Washington	3,040,000	4.7%	5.2%	
3.	Arizona	2,332,000	4.4%	3.9%	
4.	Maryland	2,768,000	3.5%	3.9%	

Directions: Use the table above to solve problem 5.

5. How many people were unemployed in Maryland in 2000? ______

Directions: Find each average annual turnover rate as a percent. Round to the nearest whole percent.

	Total Employees	Turnover for 5 Years	Turnover Rate
6.	680	62, 87, 91, 48, 52	
7.	10,358	140, 209, 321, 458, 322	
8.	11,650	2,450; 2,316; 2,607; 2,193; 2,084	
9.	47	15, 10, 10, 9, 21	
10.	2,329	548, 871, 908, 755, 643	

Employee Performance

EXAMPLE Arturo's company uses the following job performance ratings to determine employees' raises.

Adequate → 3% raise
Good → 6% raise
Outstanding → 15% raise

Arturo's supervisor rates his job performance as adequate. His hourly wage now is $15.00. What is the amount of his raise?

Multiply the current wage by the percent raise for an adequate rating, written as a decimal.

$\$15.00 \times .03 = \$.45$

Arturo's raise is $.45 per hour.

Directions: Find each raise. Use 3% for adequate, 6% for good, and 15% for outstanding.

	Employee	Wage	Performance	Raise
1.	Jessica	$7.00 per hour	outstanding	
2.	Felix	$20.00 per hour	adequate	
3.	Chen	$29,200 per year	outstanding	
4.	Paul	$16.00 per hour	good	
5.	Dolores	$48,500 per year	good	

Directions: Find each new wage with a 3% raise for adequate, 6% raise for good, and 15% raise for outstanding. Round annual wages to the nearest dollar.

	Employee	Current Wage	Performance	New Wage
6.	Yoni	$3,500 per month	good	
7.	Kyle	$52,640 per year	adequate	
8.	Max	$32.00 per hour	good	
9.	Heidi	$800 per week	outstanding	

Directions: Use the table above to solve problem 10. Round to the nearest dollar.

10. How much more would Kyle's annual salary be for an outstanding rating than for an adequate rating? ____________

Labor Issues

2000 Federal Poverty Guidelines	
Number in Family	Gross Annual Income
1	$8,350
2	$11,250
3	$14,150
4	$17,050
5	$19,950
6	$22,850

EXAMPLE Paul and Anita have 4 dependent children. Paul works full-time for $10.50 per hour. Is his income above or below the poverty level?

Step 1 Find Paul's gross annual income.

40 hours a week × 52 weeks a year = 2,080 hours a year

2,080 hours a year × $10.50 per hour = $21,840 per year

Step 2 Use the *2000 Federal Poverty Guidelines* at the right to find the poverty level for a family of six. Compare Paul's gross annual income to that amount.

$21,840 < $22,850

Paul's gross annual income is below the poverty level.

Directions: Use the table above. Write *above* or *below* to compare each income to the poverty level.

	Gross Pay	Number in Family	Above or Below Poverty Level?
1.	$1,300 per month	6	
2.	$15.00 per hour	4	
3.	$8.00 per hour	5	
4.	$1,900 per month	3	
5.	$9.00 per hour	5	

Directions: Write *Yes* or *No* to indicate whether the percent cost of living increase is greater than the rate of inflation. Round decimals to the nearest tenth of a percent.

	Annual Wage	Cost of Living Increase	Rate of Inflation	Greater Than Rate of Inflation?
6.	$26,000	$500	2.8%	
7.	$42,000	$2,300	3.5%	
8.	$51,000	$4,100	4%	
9.	$80,000	$5,000	3.2%	
10.	$62,450	$2,180	4.5%	

Name Date Period

More Labor Issues Practice

EXAMPLE In the table below, what is the difference between the average annual earnings of a female high school graduate and those of a female 2-year college graduate?

AVERAGE ANNUAL EARNINGS BY EDUCATION		
Education Level	**Men**	**Women**
9th–12th grade (no diploma)	$25,283	$17,313
High school diploma	$32,521	$21,893
2-year college degree	$39,873	$28,403
4-year college degree	$52,354	$36,555

Subtract to find the difference.

$28,403 – $21,893 = $6,510

Directions: Use the table above to solve problems 1–3.

1. Bob spends $5,000 for a 2-year college degree. How many years after high school will his total earnings be greater than his total earnings if he had only a high school diploma? __________

2. Sally spends $25,000 for a 4-year college degree. How many years after high school will her total earnings be greater than her total earnings if she had no diploma? __________

3. Sharon spends $4,000 for a 2-year college degree. How many years after high school will her total earnings be greater than her total earnings if she had no diploma? __________

Directions: Make a bar graph on the back of this page to display the data below.

	Occupation	**Median Annual Income**
4.	Air traffic controller	$73,910
5.	Animal breeder	$25,050
6.	Carpenter	$34,420
7.	Carpet installer	$31,750
8.	Database administrator	$52,550
9.	Home health aide	$18,810
10.	Pest control worker	$24,120

Name Date Period

Chapter 5, Lesson 1

Workbook 23

Comparing Fares

EXAMPLE Sid buys a round-trip ticket between Portland and El Paso for a trip in 5 weeks. He will leave on a Sunday and return on a Wednesday. What does his round-trip ticket plus tax cost on FLY Airlines?

FLY Airlines One-Way Fare (Rates do not include a 5% tax.)					
CITY		REGULAR		2-WEEK ADVANCE PURCHASE*	
From (To)	To (From)	Fri.–Mon.	Tues.–Thurs.	Fri.–Mon.	Tues.–Thurs.
Detroit	Clarksburg	$314	$254	$202	$154
Portland	El Paso	$462	$388	$350	$299
Omaha	San Diego	$395	$316	$224	$183
Boston	St. Louis	$215	$183	$68	$68
*Tickets purchased at least 14 days before departure date.					

5 weeks = $5 \times 7 = 35$ days $>$ 14 days   The ticket is advance purchase.

Step 1 Add the advance-purchase fares from Portland to El Paso on a Sunday, and from El Paso to Portland on a Wednesday.

$350 Portland to El Paso on a Sunday
+$299 El Paso to Portland on a Wednesday
$649 Round-trip airfare

Step 2 Compute the tax. Add it to the round-trip airfare.

$\$649 \times .05 = \$32.45$ $\$649 + \$32.45 = \$681.45$

Sid's round-trip ticket plus tax costs $681.45.

Directions: Use the FLY Airlines chart to find the cost of each round-trip ticket plus tax.

	Cities	Leave	Return	Advance Purchase	Airfare
1.	Omaha/San Diego	Tues.	Thurs.	Yes	
2.	Detroit/Clarksburg	Mon.	Fri.	No	
3.	Clarksburg/Detroit	Fri.	Fri.	Yes	
4.	Omaha/San Diego	Mon.	Wed.	Yes	
5.	Boston/St. Louis	Sat.	Tues.	No	
6.	San Diego/Omaha	Thurs.	Sat.	No	
7.	El Paso/Portland	Wed.	Fri.	No	
8.	St. Louis/Boston	Sun.	Tues.	Yes	
9.	Portland/El Paso	Sat.	Mon.	Yes	
10.	Clarksburg/Detroit	Tues.	Sun.	No	

Name Date Period

More Comparing Fares Practice

EXAMPLE Bernice flies from Detroit to Clarksburg on Wednesday, August 8, and returns on August 10. She buys her ticket 20 days before she leaves. Which airline offers the lower airfare before tax, FLY or Gum?

FLY Airlines/Gum Airlines One-Way Fares (Rates do not include a 5% tax.)					
City		**Regular**		**2-Week Advance Purchase**	
From (To)	**To (From)**	**Fri.–Mon.**	**Tues.–Thurs.**	**Fri.–Mon.**	**Tues.–Thurs.**
Detroit	Clarksburg	\$314/\$299	\$254/\$237	\$202/\$184	\$154/\$114
Portland	El Paso	\$462/\$511	\$388/\$426	\$350/\$375	\$299/\$308
Omaha	San Diego	\$395/\$499	\$316/\$387	\$224/\$265	\$183/\$163
Boston	St. Louis	\$215/\$234	\$183/\$176	\$68/\$84	\$68/\$45

20 days > 14 days The ticket is advance purchase.

Step 1 Find which day of the week Bernice returns. August 8 is a Wednesday, so August 9 is a Thursday, and August 10 is a Friday.

Step 2 Find the advance-purchase fares for each airline. Compare.
FLY: \$154 + \$202 = \$356 Gum: \$114 + \$184 = \$298 \$356 > \$298

Gum Airlines offers the lower airfare.

Directions: Use the airfares above to solve problems 1–10. Remember to include the tax.

1. Which airline offers the lower regular fares between Portland and El Paso? ______
2. Which airline offers the lower regular fares between Omaha and San Diego? ______
3. Which airline offers the lower advance-purchase fares between Detroit and Clarksburg? ______
4. What is the difference in price between FLY's and Gum's lowest round-trip fares between Detroit and Clarksburg? ______
5. What is the difference in price between FLY's and Gum's lowest round-trip fares between Omaha and San Diego? ______
6. What is the difference in price between FLY's and Gum's lowest round-trip fares between El Paso and Portland? ______
7. What is the difference in price between FLY's and Gum's highest round-trip fares between Portland and El Paso? ______
8. Kirk flies between Boston and St. Louis. He leaves on a Friday and returns on a Tuesday, with no advance purchase. Which airline offers the lower regular fare? How much lower? ______
9. Monica buys an advance-purchase ticket between Portland and El Paso. She returns on a Saturday. How much could she save by leaving on a Tuesday instead of a Monday, flying Gum? ______
10. Terry buys an advance-purchase ticket between Omaha and San Diego. He leaves on a Monday. How much could he save by returning on a Thursday instead of a Friday, flying FLY? ______

Selecting Lodging

EXAMPLE Salespeople on a trip need 4 doubles and a suite for 2 nights, a conference room, and Internet service. What is the least cost for lodgings that meet all the needs?

Hotel	Daily Rates	Extras
Byway Inn	Single: $79 Double: $129	None
Easterly Inn	Single: $149 Double: $229 Suite: $339	Conference room, room service, in-room fax/Internet
Your Hotel	Single: $119 Double: $149 Suite: $309	Conference room, exercise room, in-room fax/Internet

Easterly Inn and Your Hotel meet all the needs. Of these, Your Hotel has the lower rates.

Four doubles and a suite for 1 night cost ($149 × 4) + $309 = $596 + $309 = $905. The cost for 2 nights is $905 × 2 = $1,810.

The least cost is $1,810 to stay 2 nights at Your Hotel.

Directions: Use the chart to complete. Choose the hotel that meets all needs for the least cost.

	Number of Days	Rooms Needed	Special Needs	Hotel Name	Daily Rate	Total Cost
1.	4	3 doubles	Conference room			
2.	3	2 singles	None			
3.	2	1 suite	Room service			
4.	5	4 suites	Conference room			
5.	2	2 doubles	Room service, Internet			
6.	4	4 doubles	Conference room, Internet			
7.	3	1 double, 1 suite	Exercise room, fax			
8.	2	2 singles, 1 suite	Conference room, room service			
9.	4	3 singles, 4 doubles	None			

Directions: Solve problem 10.

10. The average hotel rate for a single is $196 in a large city and $84 in a small town. How much more does an average single cost for 3 nights in the city than in the town? ______

Name Date Period

Using Credit Cards

EXAMPLE Tim's statement to the right shows only business expenses that he charged to his credit card. How much did Tim charge for car rentals?

Add the car rental costs.

$163.58 + $65.82 + $291.72 = $521.12

Tim charged $521.12 for car rentals.

DATE	COMPANY NAME	LOCATION	AMOUNT
9/4	FLY Airlines	Renton, Ohio	$878.75
9/6	Ace Car Rental	Salem, Oregon	$163.58
9/6	Sam's Diner	Salem, Oregon	$31.02
9/7	Lucky Hotel	Salem, Oregon	$296.64
9/8	Airport Parking	Renton, Ohio	$42.00
9/9	FAH Gas	Renton, Ohio	$11.27
9/10	Dine In	Los Angeles, California	$109.67
9/10	We Print	Los Angeles, California	$71.58
9/10	Blue Cab	Los Angeles, California	$26.00
9/10	Beta Car Rental	Los Angeles, California	$65.82
9/10	Airport Parking	Renton, Ohio	$15.00
9/12	FLY Airlines	Renton, Ohio	$301.72
9/13	Meals Deals	San Antonio, Texas	$29.75
9/14	Shelly's Cuisine	San Antonio, Texas	$92.15
9/15	We Print	San Antonio, Texas	$288.13
9/16	Apex Car Rental	San Antonio, Texas	$291.72
9/17	Shady Hotel	San Antonio, Texas	$256.01
9/17	Airport Parking	Renton, Ohio	$35.00
9/18	FAH Gas	Renton, Ohio	$13.19

Directions: Use Tim's credit-card statement to solve problems 1–10.

1. What are Tim's charges for cab fare? ________
2. What are Tim's charges for gas? ________
3. What are Tim's charges for printing? ________
4. What are Tim's charges for airfare? ________
5. What are Tim's charges for meals? ________
6. What are Tim's charges for hotels? ________
7. What are Tim's charges for parking? ________
8. What are Tim's total business charges? ________
9. About what percentage of the total charges is for parking and printing? ________
10. About what percentage of the total charges is for airfare and meals? ________

Travel Reimbursement Forms

EXAMPLE Jenny Jones' statement and receipt show only business expenses. What is Jenny's first entry on her travel reimbursement form?

DATE	COMPANY NAME	LOCATION	AMOUNT
4/1	PJ Print	Columbus, Ohio	$106.35
4/2	Gum Airlines	Columbus, Ohio	$192.54
4/4	Jake's Place	Wichita, Kansas	$34.98
4/4	U-Rent Car Rental	Wichita, Kansas	$108.35
4/5	P&G Gas	Wichita, Kansas	$12.50
4/5	Slumber Hotel	Wichita, Kansas	$291.56
4/8	ABC Mail	Columbus, Ohio	$52.98

Air Parking
============
Date: 4/4
Amount: $18.00
transportation expense

Jenny's first business expense was on 4/1 to PJ Print for $106.35.

Jenny's first entry on her reimbursement form is dated 4/1, to PJ Print, for $106.35.

Directions: Complete the travel reimbursement form using the credit-card statement and cash receipt.

Expenses incurred by: **1.** ______________ Date: *4/28/01*

Social Security #: *000000000* Company Number: *52961*

	DATE	COMPANY	AIRFARE	LODGING	RENTAL CAR & GAS	MEALS	MISC.	TOTAL
	4/1	PJ Print					*$106.35*	*$106.35*
2.	4/2	Gum Airlines						
3.	4/2	Jake's Place						
4.	4/4	U-Rent Car Rental						
5.	4/4	Air Parking						
6.	4/5	Slumber Hotel						
7.	4/5	P&G Gas						
8.	4/8	ABC Mail						
9.		**TOTAL**						

Directions: Use the completed travel reimbursement form to solve problem 10.

10. About what percentage of Jenny's business expenses is for lodging and meals? ______________

Name Date Period

Simple Interest

EXAMPLE A company borrows $4,000 from a bank that charges 18% simple interest each year. What is the total amount the company has to pay back at the end of $4\frac{1}{2}$ years?

Step 1 Find the interest. Use the formula $I = PRT$, where $P = \$4,000$, $R = 18\%$, and $T = 4\frac{1}{2}$ years.

$I = \$4,000 \times 18\% \times 4\frac{1}{2} = \$4,000 \times .18 \times 4.5 = \$720 \times 4.5 = \$3,240$

Step 2 Add the interest to the principal. $4,000 + $3,240 = $7,240

The company has to pay back $7,240 at the end of $4\frac{1}{2}$ years.

Directions: Find the interest and amount owed at the end of each time period using the annual interest rate.

	Principal	Annual Rate	Time	Interest	Amount Owed
1.	$2,000	5%	1 year		
2.	$3,000	8%	2 years		
3.	$1,000	4%	4 years		
4.	$20,000	6%	3 years		
5.	$6,500	10%	6 years		
6.	$220,000	18%	3 months		
7.	$11,500	9%	$3\frac{1}{2}$ years		
8.	$45,000	16%	6 months		
9.	$90,000	20%	1 year, 3 months		
10.	$30,000	15%	1 year, 9 months		

Directions: Find the interest and amount owed at the end of each time period using the monthly interest rate.

	Principal	Annual Rate	Time	Interest	Amount Owed
11.	$20,400	12%	4 months		
12.	$52,000	18%	7 months		
13.	$35,000	24%	5 months		
14.	$68,200	18%	1 year, 6 months		
15.	$394,000	15%	1 year, 8 months		

Name Date Period

Compound Interest

EXAMPLE A company borrows $20,000 at an annual rate of 18% compounded monthly. How much does the company owe after 2 months?

Step 1 Find the monthly interest rate. $18\% \div 12 = 1.5\% = .015$
Use the formula $I = PRT$ to find the interest for the first month.
$I = \$20{,}000 \times .015 \times 1 = \$300 \times 1 = \$300$
Add the interest to the principal to find the amount owed after 1 month.
$\$20{,}000 + \$300 = \$20{,}300$

Step 2 Use the formula $I = PRT$ to find the interest for the second month.
$I = \$20{,}300 \times .015 \times 1 = \$304.50 \times 1 = \$304.50$
Add the interest to $20,300 to find the amount owed after the second month. Round to the nearest dollar.
$\$20{,}300 + \$304.50 = \$20{,}604.50 \approx \$20{,}605$

After 2 months the company owes $20,605.

Directions: Find the amount owed after 3 months when interest is compounded monthly. Round money to the nearest dollar.

	Principal	Annual Interest Rate	Monthly Interest Rate	Amount Owed
1.	$3,000	24%		
2.	$5,000	12%		
3.	$20,000	15%		
4.	$14,500	18%		
5.	$36,000	6%		

Directions: Find the amount owed after 1 year when interest is compounded semiannually. Round money to the nearest dollar.

	Principal	Annual Interest Rate	Semiannual Interest Rate	Amount Owed
6.	$60,000	18%		
7.	$126,000	12%		
8.	$680,000	10%		
9.	$28,000	14%		
10.	$840,000	24%		

Directions: Find the amount owed at the end of 1 year when interest is compounded quarterly. Round money to the nearest dollar.

	Principal	Annual Interest Rate	Amount Owed
11.	$2,000	16%	
12.	$8,000	20%	
13.	$4,500	12%	
14.	$25,400	24%	
15.	$40,000	8.8%	

Name Date Period

Chapter 6, Lesson 3

Workbook **30**

Business Loans

EXAMPLE A company has the assets shown in the table. Its line of credit is 65% of the total value of its assets. How much money can the company borrow using its line of credit?

Asset	Value
Inventory	\$160,000
Receivables	\$48,000
Real estate	\$320,100
Equipment	\$52,000
Cash/savings	\$18,255

Step 1 Find the total value of the assets.

$\$160{,}000 + \$48{,}000 + \$320{,}100 + \$52{,}000 + \$18{,}255 = \$598{,}355$

Step 2 Find 65% of the total assets. Round to the nearest dollar.

$\$598{,}355 \times .65 = \$388{,}930.75 \approx \$388{,}931$

The company can borrow up to \$388,931 using its line of credit.

Directions: Find the line of credit using each company's percentage of assets. Round money to the nearest dollar.

	ASSETS					Percentage of Assets	Line of Credit
	Inv.	Rec.	Real Estate	Equip.	Cash/sav.		
1.	\$50,000	\$220,000	\$645,000	\$45,000	\$13,584	60%	
2.	\$215,000	\$196,300	\$72,000	\$11,482	\$9,683	70%	
3.	\$46,258	\$71,923	\$3,162,000	\$35,882	\$21,479	55%	
4.	\$834,516	\$915,030	\$1,308,527	\$264,117	\$91,408	74%	
5.	\$2,300,000	\$706,224	\$2,025,509	\$48,232	\$193,087	62%	

Directions: Find the interest and total amount owed for each simple interest rate. Round to the nearest dollar.

	Principal	Annual Rate	Time	Interest	Amount Owed
6.	\$50,000	12%	6 months		
7.	\$260,000	9%	9 months		
8.	\$400,000	18%	5 months		

Directions: Find the amount owed after 2 months when interest is compounded monthly. Round to the nearest dollar.

	Principal	Annual Rate	Amount Owed
9.	\$70,000	12%	
10.	\$430,000	18%	

Name Date Period

Cash Flow

EXAMPLE The cash flow statement below shows the cash flow in January, and estimates the revenues and expenses for February through July. What is the cash flow in January?

	Jan.	Feb.	Mar.	Apr.	May	June	July
INCOME							
Cash Sales	\$4,293	\$3,841	\$4,255	\$3,004	\$6,792	\$7,858	\$8,592
Receivables	\$7,358	\$5,087	\$3,978	\$4,119	\$4,853	\$5,964	\$7,226
Other Income	\$652	\$463	\$659	\$853	\$1,425	\$1,853	\$2,137
Total Revenue	???						
EXPENSES							
Inventory	\$6,924	\$4,925	\$5,926	\$5,884	\$5,921	\$4,094	\$4,982
Payroll	\$3,518	\$3,446	\$3,017	\$3,149	\$3,659	\$3,259	\$3,758
Maintenance/Repairs	\$356	\$925	\$1,324	\$458	\$523	\$253	\$519
Rent	\$1,300	\$1,300	\$1,300	\$1,300	\$1,300	\$1,300	\$1,300
Marketing and Sales	\$650	\$915	\$850	\$675	\$920	\$580	\$750
Interest	\$318	\$318	\$318	\$318	\$318	\$318	\$318
Total Expenses	???						
Cash Flow	???						

Total revenue: \$4,293 + \$7,358 + \$652 = \$12,303
Total expenses: \$6,924 + \$3,518 + \$356 + \$1,300 + \$650 + \$318 = \$13,066
Subtract the lesser number from the greater number: \$13,066 – \$12,303 = \$763
Since total expenses are greater than total revenue, the cash flow is negative.

The month of January had a negative cash flow of \$763, which is a cash flow of (\$763).

Directions: Use the cash flow statement to complete the table.

	Month	Total Revenue	Total Expenses	Difference	Cash Flow
1.	February				
2.	March				
3.	April				
4.	May				
5.	June				
6.	July				

Directions: Use the cash flow statement to solve.

7. To the nearest thousand dollars, how much does this company need to borrow to meet its expenses through April? __________

8. The company borrows the amount of money in problem 7 for 2 months at 12% annual interest compounded monthly. How much money does the company owe at the end of 2 months? __________

Name Date Period

Product Payments

EXAMPLE A company's credit card sales are \$92,800 in May. It pays a 2% monthly fee on all credit transactions. How much does the company spend for credit card fees in May?

Multiply credit card sales by the percent. $\$92,800 \times 2\% = \$92,800 \times .02 = \$1,856$

The company spends \$1,856 for credit card fees.

Directions: Solve problems 1–10.

1. A company's payroll is \$192,520. The cost to process paychecks is about 1.8% of the total payroll. About how much does processing paychecks cost? ________
2. A company's payroll is \$308,680. The cost to process paychecks is about 1.5% of the total payroll. About how much does processing paychecks cost? ________
3. A company's credit card sales are \$925,800 in June. It pays a 3% monthly fee on all credit transactions. How much does the company spend for credit card fees in June? ________
4. A company spends 2% of its annual sales of \$791,500 for dishonored checks and collection fees. How much money is spent for dishonored checks and collection fees? ________
5. A company pays a checking account fee of \$15.00 a month plus \$.05 for each check it issues. If it issues 190 checks a month, how much does it spend for checking account fees that year? ________
6. A company pays a checking account fee of \$20.00 a month plus \$.09 for each check it issues. If it issues 215 checks a month, how much does it spend for checking account fees that year? ________
7. A company spends 2% of its total annual sales to process cash sales. It spends 4% of its total annual sales to process checks. If its total annual sales are \$856,110, how much more does it cost to process checks than cash sales? ________
8. A company spends 1% of its total annual sales to process cash sales. It spends 2.5% of its total annual sales to process checks. If its total annual sales are \$796,480, how much more does it cost to process checks than cash sales? ________
9. Product payment costs are 6.5% of a company's total annual expenses of \$318,400. Handling dishonored checks accounts for 1% of the company's product payment costs. How much does handling dishonored checks cost? ________
10. Product payment costs are 5.5% of a company's total annual expenses of \$268,500. Handling dishonored checks accounts for 2% of the company's product payment costs. How much does handling dishonored checks cost? ________

More Product Payments Practice

EXAMPLE One year a company spends $21,600 to process payments made by check. That year its total expenses are $483,400. About what percentage of its total annual expenses are for processing checks?

Step 1 Round the total annual expenses to the nearest $1,000.

$483,400 ≈ $483,000

Step 2 Write the ratio of the cost to the total expenses. Simplify. Divide to write the simplified ratio as a decimal, then as a percent.

$\frac{21{,}600}{483{,}000} = \frac{21{,}600 \div 100}{483{,}000 \div 100} = \frac{216}{4{,}830} = .04 = 4\%$

The company spends about 4% of its total annual expenses for processing checks.

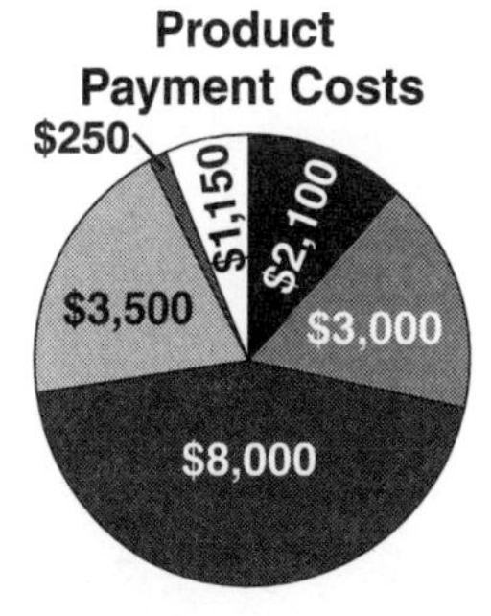

Directions: The total annual expenses of a company are $288,000. Use the circle graph above to solve. Round to the nearest hundredth of a percent.

1. Which product payment cost is the least percentage of the total product payment costs? ______________
2. About what percentage of the total annual expenses is for check processing fees? ______________
3. About what percentage of the total annual expenses is for cash transactions? ______________
4. About what percentage of the total annual expenses is for personnel costs to process checks? ______________
5. About what percentage of the total annual expenses is for dishonored check fees? ______________
6. The payroll expense is $120,000. About what percentage of payroll is the expense of personnel costs to process paychecks? ______________
7. What percentage of the total annual expenses is for product payments? ______________
8. About what percentage of product payment costs is the cost of credit transactions? ______________
9. About what percentage of product payment costs is counterfeit monies? ______________
10. Which product payment cost is about 44% of the total product payment costs? ______________

Name Date Period

Chapter 7, Lesson 1

Workbook 34

Salaries and Benefits

EXAMPLE A company's annual income is $1,800,000. It spends $900,000 annually on employee salaries. What percentage of the company's income is required for employee salaries?

Step 1 Write the ratio of the employee salaries to the company's annual income. Simplify.

$$\frac{\text{Employee salaries}}{\text{Annual income}} = \frac{\$900{,}000}{\$1{,}800{,}000} = \frac{900{,}000 \div 100{,}000}{1{,}800{,}000 \div 100{,}000} = \frac{9}{18} = \frac{1}{2}$$

Step 2 Write the fraction as a decimal. Write the decimal as a percent.

$\frac{1}{2} = 1 \div 2 = .5 = 50\%$

Paying employee salaries takes 50% of the company's income.

Directions: Find the percentage of income required for employee salaries.

	Total Income	Employee Salaries	Ratio	Percent
1.	$300,000	$120,000		
2.	$700,000	$245,000		
3.	$520,000	$338,000		
4.	$880,000	$303,600		
5.	$3,900,000	$1,891,500		

Directions: Use the bar graph at the right to solve problems 6–10.

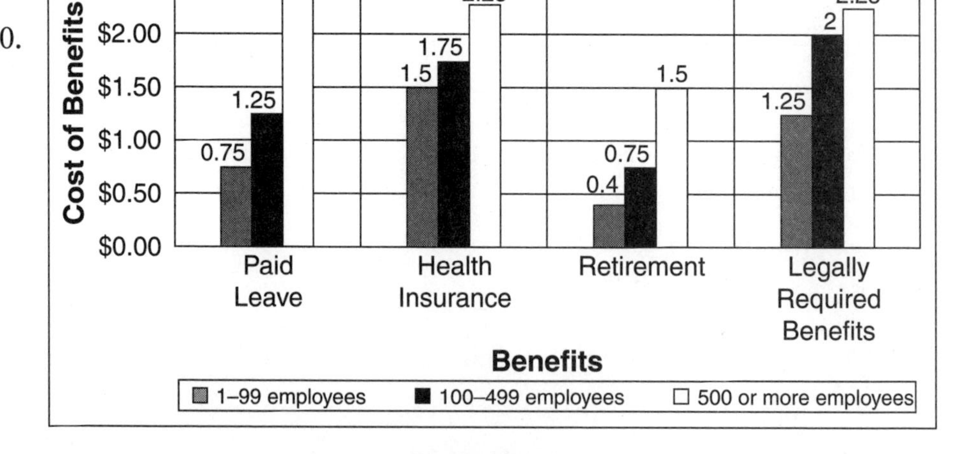

6. How much does a large company with 600 employees spend per hour worked for employee benefits? ________

7. A company has 268 employees. How much more does it spend per hour worked for paid leave than a company with 89 employees? ________

8. A company has 183 employees who work a total of 350,000 hours annually. How much does it spend for retirement benefits? ________

9. A company has 29 employees. How much more does it spend per hour worked for voluntary benefits than for legally required benefits? ________

10. A company has 520 employees who work a total of 1,300,000 hours annually. How much does it spend for paid leave and health insurance? ________

Name Date Period

Building Facilities

EXAMPLE A company makes a table of the office space it needs. How many square feet of office space does it need?

	Senior Managers	Directors	Managers	Staff	Lunchroom	Supply Room
Area per person (ft.^{2})	520	350	300	100	15	18
Number of people	2	4	8	45		

Step 1 Add to find the total number of employees. $2 + 4 + 8 + 45 = 59$

Step 2 Multiply the area per person by the number of people using the area.

$520 \times 2 = 1{,}040$ $350 \times 4 = 1{,}400$ $300 \times 8 = 2{,}400$
$100 \times 45 = 4{,}500$ $15 \times 59 = 885$ $18 \times 59 = 1{,}062$

Step 3 Add the areas. $1{,}040 + 1{,}400 + 2{,}400 + 4{,}500 + 885 + 1{,}062 = 11{,}287$

The company needs 11,287 square feet of office space.

Directions: Find the area of office space in square feet that each company needs.

	Senior Managers		Directors		Managers		Staff		Lunchroom (area per employee)	Other Common Areas (area per employee)	Total Square Feet Needed
	Area	No.	Area	No.	Area	No.	Area	No.			
1.	700	1	600	5	550	8	200	45	16	20	
2.	800	1	600	4	480	9	300	52	14	18	
3.	650	3	480	4	400	10	220	80	25	22	
4.	680	2	500	3	380	9	85	65	18	30	
5.	750	4	550	5	410	10	55	78	12	32	

Directions: Solve problems 6–10.

6. A company buys 8 new computers for $1,919 each and 6 software programs for $409 each. How much do the computers and software cost? ______

7. A company buys 35 new computers for $2,389 each and 35 software programs for $224 each. How much do the computers and software cost? ______

8. A Banko computer sells for $1,535 and a Crunch computer sells for $2,629. How much money does a company save by buying 20 Banko computers instead of 20 Crunch computers? ______

9. A company buys 12 computers for $2,105 each and 10 software programs for $1,195 each. Each computer has a $300 rebate. How much do the computers and software cost after the rebates? ______

10. A company buys 16 computers for $3,008 each, 12 computers for $2,959 each, and 20 software programs for $1,595 each. Each computer has a $150 rebate. How much do the computers and software cost after the rebates? ______

Name Date Period

Chapter 7, Lesson 3

Workbook 36

Transportation

EXAMPLE A company's sales force uses 9 company cars. The table shows the monthly costs of maintaining and operating these cars. Find the company's total monthly transportation expense.

Average Monthly Mileage	Gas Cost (per mile)	Maintenance/Repair (per mile)	Insurance Cost (per car)	Depreciation Cost (per mile)
8,000	$.06	$.28	$352	$.35

Step 1 Find the cost for each category.

Gas
$\$.06 \times 8{,}000 = \$480$

Maintenance/Repair
$\$.28 \times 8{,}000 = \$2{,}240$

Insurance
$\$352 \times 9 = \$3{,}168$

Depreciation
$\$.35 \times 8{,}000 = \$2{,}800$

Step 2 Add all the costs. $\$480 + \$2{,}240 + \$3{,}168 + \$2{,}800 = \$8{,}688$

The company's monthly transportation expense is $8,688.

Directions: Find each monthly transportation expense.

	Number of Cars	Average Monthly Mileage	Gas (per mile)	Maintenance/ Repair (per mile)	Insurance (per car)	Depreciation (per mile)	Monthly Transportation Expense
1.	25	48,000	$.09	$.27	$215	$.51	
2.	15	29,000	$.12	$.35	$345	$.29	
3.	35	71,000	$.10	$.31	$185	$.38	
4.	11	18,500	$.08	$.26	$368	$.52	
5.	156	220,000	$.05	$.32	$55	$.45	

Directions: Find each company's transportation costs.

	Company	Fraction of Operating Expenses for Transportation	Operating Expenses Budget	Transportation Budget
6.	Richard's Real Estate	$\frac{1}{3}$	$55,140	
7.	Concrete Mixers	$\frac{1}{2}$	$468,000	
8.	Danny's Deli	$\frac{1}{10}$	$68,350	
9.	Sam's Cleaners	$\frac{3}{5}$	$42,950	
10.	Idaho Trucking	$\frac{7}{8}$	$652,160	

Name Date Period

Cost of Production

EXAMPLE The line graph to the right shows three relationships: between output and fixed costs, between output and variable costs, and between output and total costs. What is the fixed cost to produce 500 units?

Find 500 on the "Output" axis.

Move directly up to the *Fixed Costs* line. Look directly to the left for the value on the "Production Cost" axis. The production cost on the variable costs graph is $400 for an output of 500.

The fixed cost is $400 to produce 500 units.

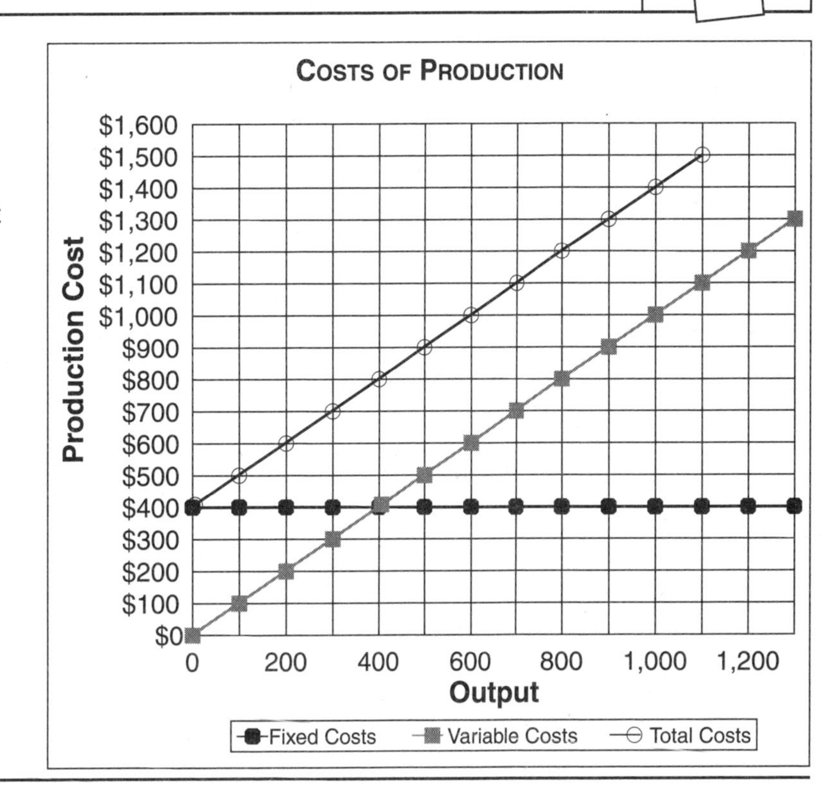

Directions: Use the graph above for problems 1–10.

1. What is the fixed cost to produce 600 units? __________
2. What is the variable cost to produce 200 units? __________
3. What is the variable cost to produce 800 units? the total cost to produce 800 units? __________
4. What is the fixed cost to produce 700 units? the total cost to produce 700 units? __________
5. How much more is the variable cost to produce 900 units than the variable cost to produce 300 units? __________
6. How much more is the variable cost to produce 800 units than the variable cost to produce 400 units? __________
7. How much more is the total cost to produce 1,000 units than the total cost to produce 500 units? __________
8. How much more is the total cost to produce 600 units than the total cost to produce 300 units? __________
9. What is the total cost to produce 1,200 units? Use the graph of fixed costs and variable costs. __________
10. What is the total cost to produce 1,300 units? Use the graph of fixed costs and variable costs. __________

More Cost of Production Practice

EXAMPLE Make a vertical double bar graph to show K & B's fixed and variable costs for July through December.

K & B Sport Ball Manufacturers Production Costs, 7/02–12/02						
Costs	**July**	**August**	**September**	**October**	**November**	**December**
Fixed	$6,000	$6,000	$6,000	$6,000	$6,000	$6,000
Variable	$15,000	$10,000	$5,000	$10,000	$10,000	$15,000

Choose a scale. Mark off the vertical axis in $2,000 units from $0 to $16,000.

Label the vertical and horizontal axes.

Draw bars to represent the data. Use two different colors.

Give the graph a title.

Make a key that shows what the two different colors represent.

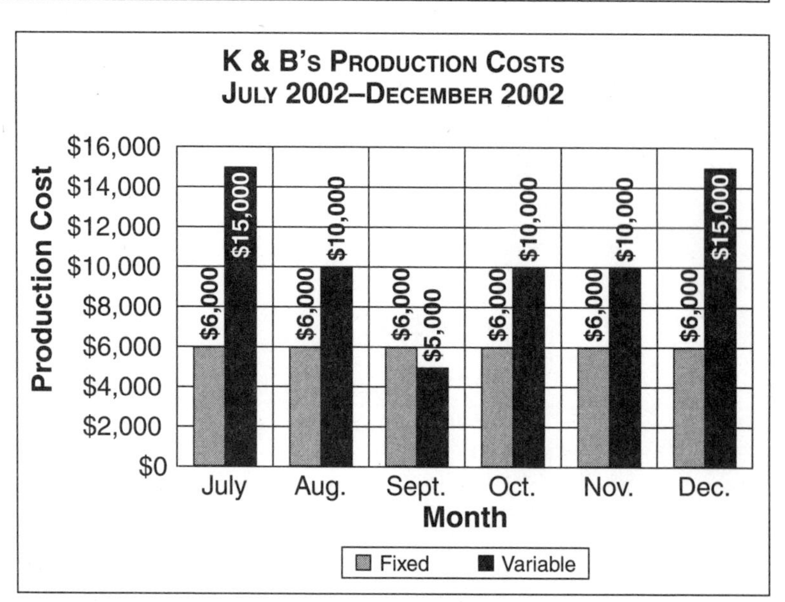

Directions: Make a horizontal double bar graph on the back of this page to show the fixed and variable production costs.

1. Tony's Noodle Manufacturing Production Costs, 6/03–11/03						
Costs	**June**	**July**	**August**	**September**	**October**	**November**
Fixed	$2,000	$2,000	$2,000	$2,000	$2,000	$2,000
Variable	$15,000	$5,000	$16,000	$20,000	$15,000	$30,000

2. O & Y Manufacturing Production Costs, 1/04–6/04						
Costs	**January**	**February**	**March**	**April**	**May**	**June**
Fixed	$1,000	$1,000	$1,000	$1,000	$1,000	$1,000
Variable	$3,000	$6,000	$4,000	$10,000	$8,000	$16,000

Name Date Period

Cost of Sales

EXAMPLE The table below shows the marketing expenses and sales for the Block Company. Total marketing expenses are what percentage of total sales for January?

Step 1 The total marketing expenses are $275,000 + $450,000 + $92,000 + $118,300 + $38,175 + $82,100 + $26,230 = $1,081,805.

Step 2 Divide total marketing expenses by total sales. Round to the thousandths place. Write the decimal as a percent.

$1,081,805 ÷ $42,900,000 ≈ .025 = 2.5%

The total marketing expenses are 2.5% of Block Company's total January sales.

Directions: Complete the table. Then use the table for problems 14–15. Round to the nearest tenth of a percent.

BLOCK COMPANY FIRST QUARTER, 2003				
Marketing Expense	**January**	**February**	**March**	**Totals**
TV advertising	$275,000	$275,000	$275,000	**1.**
Newspaper/magazine advertising	$450,000	$450,000	$450,000	**2.**
Radio advertising	$92,000	$92,000	$92,000	**3.**
Web sites	$118,300	$91,300	$123,200	**4.**
Promotions	$38,175	$24,250	$13,900	**5.**
Market research	$82,100	$67,400	$54,200	**6.**
Other	$26,230	$31,020	$28,250	**7.**
Total marketing expenses	?????	**8.**	**9.**	**10.**
Total Sales	$42,900,000	$51,300,000	$58,100,000	$152,300,000
Percentage of sales for marketing	?????	**11.**	**12.**	**13.**

14. In which month did Block Company spend the greatest percentage of sales for marketing expenses? the least percentage? __________

15. What was the average monthly marketing expense? __________

Name Date Period

Inventory

EXAMPLE The table below shows the number and cost of stair climbers purchased by a fitness store during May. At the end of May, 35 stair climbers are in stock. What is the value of the inventory on May 31?

Date	Inventory	Unit Cost	Total Cost
Beginning inventory	30	$286	$8,580
May 8 purchase	16	$318	$5,088
May 16 purchase	24	$204	$4,896
Totals	70		$18,564

Step 1 Find the average cost. $18,564 ÷ 70 = $265.20

Step 2 Multiply the average cost by the number of stair climbers in inventory on May 31.
$265.20 × 35 = $9,282

The value of the inventory on May 31 is $9,282.

Directions: Complete the table. Find the average cost. Find the value of the inventory at the end of the month, when 48 items are in stock. Round to the nearest cent.

Date	Inventory	Unit Cost	Total Cost
Beginning inventory	25	$62	**1.**
July 10 purchase	34	$58	**2.**
July 20 purchase	28	$105	**3.**
Totals	87		$6,462

Average Cost of Inventory: **4.** ____________

Value of Inventory: **5.** ____________

Directions: Find the average inventory value.

	Beginning Inventory	Ending Inventory	Average Inventory
6.	$8,200	$12,800	
7.	$38,100	$24,640	

Directions: Find the inventory turnover. Round to the nearest tenth.

	Average Inventory	Cost of Goods Sold	Inventory Turnover
8.	$52,400	$183,400	
9.	$219,500	$691,000	

Directions: Solve problem 10. Round to the nearest tenth.

10. The inventory at Morton's Mirrors was $4,120 on March 1 and $2,570 on March 31. The cost of goods sold in March was $8,110. What was the inventory turnover? ____________

Name Date Period

Profit and Loss

EXAMPLE The Cooke Company's annual income statements for the last two years are shown below. Find the net sales, gross profit, and net income before income tax for 2001. Find the percent increase in sales from 2001 to 2002.

Step 1 Net sales = Sales − Returns = $350,000 − $19,100 = $330,900

Gross profit = Net sales − Cost of goods sold = $330,900 − $87,200 = $243,700

Net income = Gross profit − Operating expenses = $243,700 − $23,523 = $220,177

Step 2 Subtract to find the change in sales. $606,995 − $330,900 = $276,095

Step 3 Divide the change by the sales in 2001, the base year. Round to two decimal places. Write the decimal as a percent.

$276,095 ÷ $330,900 ≈ .83 = 83%

Sales increased about 83% between 2001 and 2002.

Directions: Complete the table. Round to the nearest whole percent.

The Cooke Company
Annual Income Statements for 2001 and 2002

	2001	2002	Change	Percent Change
REVENUE FROM SALES:				
Sales	$350,000	$628,300	$278,300	80%
Less sales returns and allowances	19,100	21,305	$ 2,205	**1.** _____
Net sales	?????	$606,995	?????	?????
COST OF GOODS SOLD:				
Inventory 1/1	$ 91,300	$ 95,500	**2.**	**3.**
Plus purchases	58,700	76,100	**4.** _____	**5.** _____
Merchandise available for sale	$150,000	**6.**	**7.**	**8.**
Less inventory 12/31	62,800	68,400	$ 5,600	9%
Cost of goods sold	$ 87,200	**9.**	**10.**	**11.**
Gross profit	????	**12.**	**13.**	**14.**
OPERATING EXPENSES:				
Cost of sales	$ 13,200	$ 21,425	**15.**	**16.**
General expenses	6,108	5,580	**17.**	**18.**
Interest expense	4,215	6,250	**19.** _____	**20.** _____
Total operating expenses	$23,523	$ 33,255	**21.**	**22.**
Net income before income tax	????	**23.**	**24.**	**25.**

Name Date Period

Kinds of Insurance

EXAMPLE Each month Batch Industries pays for these casualty insurance policies: fire/theft, $2,815; liability, $3,104; automobile, $1,012.40; and Workers' Compensation, $2,857.62. What is the total annual premium for casualty insurance?

Step 1 Find the total monthly premium.
$2,815 + $3,104 + $1,012.40 + $2,857.62 = $9,789.02

Step 2 Multiply by 12 to find the annual premium.
$9,789.02 × 12 = $117,468.24

The total annual premium for casualty insurance is $117,468.24.

Directions: Find the total monthly and annual premiums.

	Company	Fire/ Theft	General Liability	Auto	Workers' Compensation	Total Monthly Premium	Total Annual Premium
1.	A	$2,023.46	$2,382.02	$396.25	$2,235.62		
2.	B	$2,952.31	$2,245.99	$413.18	$2,235.62		
3.	C	$2,817.16	$1,892.17	$358.43	$2,235.62		
4.	D	$2,325.94	$2,256.14	$443.06	$2,235.62		
5.	E	$2,608.22	$1,962.35	$402.11	$2,235.62		

Directions: Use the table below and the formula $A = lw$ to solve problems 6–8.

Area	Annual Premium
Less than 2,000 square feet	$800
2,000 square feet to 3,999 square feet	$1,500
4,000 square feet to 5,999 square feet	$2,800

6. Bert's Hardware has 1,852 square feet of floor space. How much is its annual liability insurance premium? ____________

7. Tot's Toys has floor space measuring 108 feet by 42 feet. How much is its annual liability insurance premium? ____________

8. Annette's Gifts has floor space measuring 54 feet by 65 feet. How much is its annual liability insurance premium? ____________

Directions: The given premiums are annual. Solve problems 9–10.

9. A company has 26 trucks. The total premium for 1 truck is $795. What is the total premium for 26 trucks with a 10% discount? ____________

10. A store has 18 trucks. The premiums for 1 truck are: bodily injury, $192; property damage, $142; comprehensive, $295; collision, $253; and medical, $58. What is the total premium with a 15% discount? ____________

Name Date Period

Cost of Insurance

EXAMPLE A clothing manufacturer buys fire/theft insurance. The annual cost is $10.75 per $10,000 of coverage. The company needs $300,000 in coverage. What is the annual premium?

Step 1 Write a proportion. $\frac{\text{Cost of coverage}}{\text{Amount of coverage}}$ $\frac{10.75}{10{,}000} = \frac{■}{300{,}000}$

Step 2 Use the cross products to solve.

$$10.75 \times 300{,}000 = 10{,}000 \times ■$$
$$3{,}225{,}000 = 10{,}000■$$
$$3{,}225{,}000 \div 10{,}000 = ■$$
$$322.5 = ■$$

The annual premium for fire/theft insurance is $322.50.

Directions: Find the annual premium for each amount of coverage.

	Coverage	Cost per $10,000 of Coverage	Annual Premium
1.	$800,000	$18.25	
2.	$500,000	$12.34	
3.	$700,000	$16.54	
4.	$200,000	$8.25	
5.	$100,000	$6.92	
6.	$900,000	$22.46	
7.	$600,000	$14.38	
8.	$1,300,000	$23.18	
9.	$3,000,000	$21.08	
10.	$1,600,000	$15.20	
11.	$1,400,000	$19.64	
12.	$2,800,000	$18.55	
13.	$4,000,000	$21.06	
14.	$3,600,000	$23.41	
15.	$4,800,000	$29.04	

Name Date Period

Chapter 9, Lesson 2

Workbook **44**

More Cost of Insurance Practice

EXAMPLE Murlow Industries owns 12 vehicles. Its coverage is 50/100 for bodily injury, 50 for property damage, and $4,000 for medical. It has a $100 deductible for comprehensive and collision. What is the company's premium for all 12 vehicles?

Bodily Injury	15/30	25/50	50/100	100/300
Premium	$63	$146	$183	$218
Property Damage	10	15	50	100
Premium	$54	$125	$168	$192
Comprehensive Deductible	$100	$250	$500	$1,000
Premium	$450	$348	$210	$143
Collision Deductible	$100	$250	$500	$1,000
Premium	$485	$395	$305	$201
Medical	$1,000	$2,000	$3,000	$4,000
Premium	$16	$32	$41	$58

Step 1 Find each coverage in the table to the right. Add the premiums.

$183 + $168 + $58 + $450 + $485 = $1,344

Step 2 Multiply the premium for 1 vehicle by the number of vehicles.

$1,344 × 12 = $16,128

Murlow Industries' premium is $16,128.

The table above shows the premiums for four levels of coverage.

Directions: Find the total annual premium using the table above.

	Number of Vehicles	Bodily Injury	Property Damage	Comprehensive Deductible	Collision Deductible	Medical	Total Premium
1.	6	15/30	10	$100	$100	$1,000	
2.	15	100/300	100	$500	$500	$3,000	
3.	8	50/100	50	$250	$250	$2,000	
4.	9	15/30	10	$1,000	$1,000	$1,000	
5.	25	50/100	50	$250	$250	$3,000	
6.	12	25/50	15	$250	$250	$4,000	
7.	10	25/50	15	$100	$100	$4,000	
8.	30	15/30	10	$1,000	$1,000	$2,000	

Directions: Solve problems 9–10.

9. A company pays a premium of $9,760. It adds a rider covering towing for $\frac{3}{4}$% of the premium. What does the rider cost? ________

10. The Duck Company pays a premium of $26,200. It adds a rider covering towing for $\frac{5}{8}$% of the premium. What is the total premium for auto insurance? ________

Insurance as an Investment

EXAMPLE Luke buys a whole life policy at age 30. What is his actual cost per $1,000 coverage at age 40? His death payment at age 40 is $100,000. The cash value at age 40 is $12,350. His annual premium is $1,400.

Step 1 Death payment – Cash value = Cost for insurance company

$100,000 – $12,350 = $87,650

Step 2 Find the insurance company's cost per $1,000 of coverage.

$87,650 ÷ 1,000 = $87.65

Step 3 Divide the annual premium by the insurance company's cost per $1,000 of coverage. Round to the nearest cent.

$1,400 ÷ $87.65 ≈ $15.97

Luke's actual cost per $1,000 coverage at age 40 is about $15.97.

Directions: Find the insurance company's cost for the death payment and Luke's actual cost per $1,000 coverage for each given age. Round to the nearest cent.

	Age	Annual Premium	Death Payment	Cash Value	Cost to Insurance Co.	Actual Cost per $1,000 Coverage
1.	45	$1,400	$100,000	$15,850		
2.	50	$1,400	$100,000	$22,240		
3.	55	$1,400	$100,000	$31,200		
4.	60	$1,400	$100,000	$46,500		
5.	65	$1,400	$100,000	$52,280		

Directions: Find the total premiums paid over the term of each endowment policy.

	Policy Term (years)	Monthly Premiums	Death Payment	Cash Value at Term	Total Premiums
6.	18	$19.52	$50,000	$13,018	
7.	20	$112.58	$100,000	$62,495	
8.	20	$41.52	$350,000	$21,963	
9.	25	$62.18	$300,000	$42,005	
10.	25	$74.29	$200,000	$51,318	

Name Date Period

Environmental Regulations

EXAMPLE A company installs an FGD system on 3 smokestacks to reduce pollution. Each FGD system costs $52,058 plus $14,968 for installation. What is the total cost?

Step 1 Find the cost of one system. $\$52{,}058 + \$14{,}968 = \$67{,}026$

Step 2 Multiply to find the total cost of 3 systems. $\$67{,}026 \times 3 = \$201{,}078$

The company pays $201,078 for the 3 FGD systems.

The table shows the cost of buying and installing three types of FGD systems.

FGD Type	Cost	Installation Fee
I	$28,326	$9,018
II	$26,092	$13,549
III	$19,847	$7,348

Directions: Use the table to solve problems 1–3.

1. Eastern Electric has 6 smokestacks that require a Type III FGD system. What is the total cost to buy and install the 6 systems? __________
2. A manufacturer installs 7 Type II FGD systems and 3 Type I FGD systems. What is the total cost to buy and install the 10 systems? __________
3. An electric company may install 10 Type I FGD systems or 10 Type II FGD systems. Which type of system costs less with installation? How much less? __________

Directions: Find the total cost of the air-purifying systems.

	Number of Systems	Collection Unit	Filter	Fan	Total Cost
4.	8	$2,546.23	$2,131.41	$4,989.64	
5.	5	$2,438.16	$5,925.65	$3,784.48	
6.	12	$1,967.87	$4,904.33	$4,906.93	
7.	20	$4,102.20	$2,299.67	$3,815.45	
8.	16	$2,965.38	$3,005.85	$8,746.95	

Directions: Solve problems 9–10.

9. Metal Shavings produces 1.5 cubic yards of air pollutants monthly. Its air purifier removes 99.7% of air pollutants. How many cubic yards of pollutants are left in the air each month? __________
10. A manufacturer produces .80 cubic yards of air pollutants each month. Its air purifier collects only 93.5% of air pollutants. How many cubic yards of pollutants are left in the air each month? __________

Americans with Disabilities Act

EXAMPLE A business builds a solid concrete ramp that is 52 feet long, 3.5 feet wide, and 6 inches high. Concrete costs $65 per cubic yard. How much does the concrete for the ramp cost?

Step 1 Use the formula $V = \frac{lwh}{2}$ to find the volume of the ramp.

$l = 52$ feet $w = 3.5$ feet $h = 6$ inches $= .5$ foot

$V = \frac{52 \text{ ft.} \times 3.5 \text{ ft.} \times .5 \text{ ft.}}{2} = \frac{91 \text{ cu. ft.}}{2} = 45.5$ cu. ft.

Step 2 Since 27 cubic feet = 1 cubic yard, divide by 27 to change cubic feet to cubic yards. Round up. 45.5 cu. ft. ÷ 27 cu. ft. ≈ 2 cu. yd.

Step 3 Multiply the number of cubic yards by the cost per cubic yard.

2 cu. yd. × $65 per cu. yd. = $130

The concrete for the ramp costs $130.

Directions: Find the cost of each ramp. Remember to round up to the next cubic yard.

	Length	Width	Height	Cost of Concrete	Total Cost
1.	30 ft.	4 ft.	1 ft.	$65	
2.	49 ft.	3 ft.	1 ft.	$70	
3.	58 ft.	3.5 ft.	1.5 ft.	$75	
4.	84 ft.	4 ft.	1.5 ft.	$68	
5.	112.5 ft.	3.5 ft.	6 in.	$72	

Directions: Find the total cost of the handrails. Posts cost $5.45 per foot. Post caps cost $.95 each. Railing costs $3.39 per foot. The welder earns $45.70 per hour.

	Posts Needed	Number of Post Caps	Amount of Railing	Number of Welder's Hours	Total Cost of Handrails
6.	Twelve 6-ft. posts	12	160 ft.	15	
7.	Forty 5-ft. posts	40	304 ft.	32	
8.	Twenty 6-ft. posts	20	288 ft.	29.5	

Directions: Solve problems 9–10. Remember to round up to the next square foot.

9. A business installs 2 automatic doors with crash bars. Each door is 7 feet, 3 inches tall and $48\frac{1}{2}$ inches wide. The doors cost $52.95 per square foot. How much do both doors cost? __________

10. A business installs 2 automatic doors with crash bars. Each door is 7 feet, 6 inches tall and $47\frac{1}{2}$ inches wide. The doors cost $68.75 per square foot. How much do both doors cost? __________

Name Date Period

Interstate Commerce

EXAMPLE A trucker drove 18,500 miles in Massachusetts and 22,000 miles in New York. Massachusetts charges $.21 per gallon and New York charges $.2945 per gallon in fuel taxes. The truck gets 5.5 miles per gallon. What is the fuel tax for both states?

Step 1 Find the number of gallons used in each state. Round to the nearest whole gallon.

Massachusetts: $\frac{18{,}500 \text{ mi.}}{\blacksquare \text{ gal.}} = \frac{5.5 \text{ mi.}}{1 \text{ gal.}}$ $5.5\blacksquare = 18{,}500$ $\blacksquare = 18{,}500 \div 5.5 \approx 3{,}364$

New York: $\frac{22{,}000 \text{ mi.}}{\blacksquare \text{ gal.}} = \frac{5.5 \text{ mi.}}{1 \text{ gal.}}$ $5.5\blacksquare = 22{,}000$ $\blacksquare = 22{,}000 \div 5.5 = 4{,}000$

Step 2 Multiply the gallons used in each state by the state's rate. Add.

Massachusetts: $3{,}364 \times \$.21 = \$706.44$ New York: $4{,}000 \times \$.2945 = \$1{,}178.00$

$\$706.44 + \$1{,}178.00 = \$1{,}884.44$

The fuel tax for both states is $1,884.44.

The table shows the fuel tax rate for selected states.

State	Cost per Gallon	State	Cost per Gallon	State	Cost per Gallon
Idaho	$.25	North Dakota	$.21	Utah	$.245
Montana	$.2775	South Dakota	$.22	Wyoming	$.13

Directions: Use the table above to solve problems 1–2. Round to the nearest whole gallon before finding the tax. Round to the nearest cent.

1. A trucking company drove 32,648 miles in Montana, 9,518 miles in Wyoming, and 27,962 miles in North Dakota. The trucks get 5 miles per gallon. How much is the fuel tax? __________

2. Trucks Ahead drove 11,315 miles in Idaho and 46,397 miles in Utah. The trucks get 3.5 miles per gallon. How much is the fuel tax? __________

Directions: Find each state's mileage percent and fee. Round to the nearest cent.

	State	Total Annual Mileage	Annual State Mileage	State Mileage Percent	Apportioned Plate Fee	Fee Paid to State
3.	Nebraska	65,025	26,010		$5,903.12	
4.	Florida	72,348	25,322		$6,459.28	
5.	California	68,945	27,951		$3,421.59	
6.	Arizona	49,356	3,068		$4,919.83	
7.	Texas	128,950	72,582		$3,187.37	
8.	Iowa	92,958	31,007		$2,192.35	
9.	Mississippi	88,465	21,994		$3,343.13	
10.	Kentucky	78,681	62,070		$1,959.84	

International Business

EXAMPLE A company ships 80,000 aprons to France. The average cost of each apron is $.68. The insurance is $2,105.68 and freight is $2,599.64. What is the CIF for the shipment?

Step 1 Multiply to find the total cost of the shipment. $80{,}000 \times \$.68 = \$54{,}400$

Step 2 Add the costs. $\$54{,}400 + \$2{,}105.68 + \$2{,}599.64 = \$59{,}105.32$

The CIF for the shipment is $59,105.32.

Directions: Find the CIF for each shipment.

	Item	Number of Items	Average Cost per Item	Insurance	Freight	CIF
1.	Shoes	200,000	$25.17	$4,968.58	$12,645.83	
2.	Baseballs	900,000	$1.35	$5,628.13	$8,552.25	
3.	Mats	48,000	$5.29	$3,202.92	$6,889.97	
4.	Cabinets	24,000	$32.92	$4,191.64	$20,586.37	
5.	Desks	250,000	$92.84	$23,546.67	$35,962.31	

Directions: Compute the tariff for each CIF. Round to the nearest cent.

	CIF	Tariff Percent	Tariff
6.	$205,683	25%	
7.	$495,998	45%	
8.	$856,432	8%	

Directions: Solve problems 9–10. Round answers to the nearest whole percent.

9. A company sells a bike in its own country for $52.19. After foreign taxes and tariffs, the price is $375.86. What is the percent increase in price? ____________

10. A company sells a clock in its own country for $32.16. After foreign taxes and tariffs, the price is $192.43. What is the percent increase in price? ____________

Financial Risks

EXAMPLE Doreen owns a saddlery. She charges $52.89 for saddle pads. This year her costs increased. She decides to raise her prices by 25%. What will Doreen charge for a saddle pad?

Step 1 Multiply the price by the percent increase to find the price increase.
$\$52.89 \times 25\% = \$52.89 \times .25 = \$13.2225$

Step 2 Round the price increase to the nearest cent. Add it to the original price.
$\$13.2225 \approx \$13.22$ $\$52.89 + \$13.22 = \$66.11$

Doreen will charge $66.11 for a saddle pad.

Directions: Find the new price of each item. Round to the nearest cent.

	Item	Original Price	Percent Increase	New Price
1.	Halter	$25.75	5%	
2.	Harness	$19.95	10%	
3.	Pants	$35.85	20%	
4.	Boots	$125.00	25%	
5.	Feed bag	$16.45	15%	

Directions: Each business owner raises prices the same percent as the inflation rate. Solve. Round to the nearest cent.

6. Sam owns a beauty salon. He charges $18 for a haircut. The inflation rate is 2.9%. How much does he raise the price of a haircut? ______________

7. A heavy-duty battery charger costs $182.50. The inflation rate is 1.7%. What is the price of the battery charger when it is adjusted for inflation? ______________

Directions: Find the CPI for the same item in each city. Round to the nearest unit.

	City	Base Period		Current Period	
		Average Price	Index	Average Price	Index
8.	A	$1.08	100	$1.38	
9.	B	$2.64	100	$2.71	
10.	C	$5.93	100	$6.15	
11.	D	$.75	100	$.82	
12.	E	$.23	100	$.28	
13.	F	$29.68	100	$31.05	
14.	G	$43.35	100	$51.99	
15.	H	$18.75	100	$19.00	

Legal Risks

EXAMPLE Mr. Gomez's catastrophic insurance guarantees payment for 50% of the average crop yield. The last 4 years he has had yields of 62 bushels, 64 bushels, 63 bushels, and 58 bushels. For how many bushels per acre does his insurance guarantee payment?

Step 1 Find the average crop yield. $62 + 64 + 63 + 58 = 247$ $247 \div 4 = 61.75$

Step 2 Multiply the average yield by 50%. $61.75 \times 50\% = 61.75 \times .5 = 30.875$

Mr. Gomez's catastrophic insurance guarantees payment for 30.875 bushels per acre.

Directions: An insurance company guarantees payment for 50% of the average yield. Find each average yield and the guaranteed payment number of bushels per acre.

	Bushels per Acre for 4 Years	Average Yield per Acre	Guaranteed Payment Number of Bushels per Acre
1.	92 bushels, 88 bushels, 88 bushels, 72 bushels		
2.	101 bushels, 107 bushels, 104 bushels, 112 bushels		
3.	70 bushels, 80 bushels, 85 bushels, 75 bushels		
4.	55 bushels, 59 bushels, 61 bushels, 64 bushels		
5.	118 bushels, 125 bushels, 124 bushels, 123 bushels		

Directions: Catastrophic insurance pays 60% of the market price. Find the catastrophic insurance payment per bushel and per acre. Then find the total insurance payment for each size farm.

	Number of Farm Acres	Guaranteed Payment Number of Bushels per Acre	Bushels per Acre Produced	Market Price per Bushel	Insurance Payment per Bushel	Insurance Payment per Acre	Total Insurance Payment
6.	4,000	50	25	$4.10			
7.	3,000	80	30	$2.50			
8.	7,000	62	19	$1.40			
9.	6,809	85	27	$3.20			

Directions: Solve problem 10. Round to the nearest cent.

10. Dr. Livingston sees 840 patients a year. His malpractice insurance premiums increase $9,120. He passes on this increased cost to his patients. How much does he increase the cost of a patient visit? ____________

Name Date Period

Sales and Revenue Projections

EXAMPLE Bell Ringers Company sells doorbells that cost $8.75 for $19.95. The company's monthly operating expenses are $31,360. How many doorbells must be sold monthly to cover the operating expenses?

Step 1 Subtract the cost of goods sold from the selling price to find the gross profit per doorbell.

$19.95 – $8.75 = $11.20

Step 2 Divide the operating expenses by the gross profit per doorbell.

$31,360 ÷ $11.20 = 2,800

Bell Ringers Company must sell 2,800 doorbells each month to cover operating expenses.

Directions: Find each company's gross profit per item. Then find the number of sales needed to pay the cost of the goods sold and operating expenses.

	Company	Monthly Operating Expenses	Selling Price per Item	Cost per Item	Gross Profit per Item	Number of Sales
1.	Mighty Masks	$6,650.00	$7.35	$3.85		
2.	Orange Co.	$13,356.00	$6.15	$1.95		
3.	Happy Times	$29,326.00	$11.73	$4.91		
4.	Eye Sights	$223,046.00	$14.60	$5.44		
5.	First Baby	$367,637.76	$15.00	$3.48		

Directions: Each company sells just enough to cover monthly expenses and have 10% for future growth. Complete the table to find the number of items sold. Round the number of sales up to the next whole number.

	Company	Monthly Operating Expenses	Operating Costs plus 10% Growth	Selling Price per Item	Cost per Item Sold	Gross Profit per Item	Number of Items Sold
6.	A	$18,460		$15.25	$3.95		
7.	B	$22,280		$21.95	$11.13		
8.	C	$32,400		$35.99	$16.79		
9.	D	$26,100		$42.65	$18.34		
10.	E	$19,007		$18.72	$5.61		

Conventions and Exhibits

EXAMPLE Miko has 2,700 inquiries about its products, with 858 of the inquiries resulting in sales. What percent of the inquiries result in sales? How many inquiries are needed for 1,000 sales?

Step 1 Write the ratio of sales to inquiries as a decimal. Round to two decimal places. Write the decimal as a percent.

$\frac{858 \text{ Sales}}{2{,}700 \text{ Inquiries}}$ $858 \div 2{,}700 \approx .32 = 32\%$ About 32% of the inquiries result in sales.

Step 2 Write and solve a proportion. Round to the nearest whole number. $\frac{858}{2{,}700} = \frac{1{,}000}{■}$

$858 \times ■ = 1{,}000 \times 2{,}700$ $858■ = 2{,}700{,}000$ $■ = 2{,}700{,}000 \div 858 \approx 3{,}147$

Miko needs 3,147 inquiries to make 1,000 sales.

Directions: Solve problems 1–5. Round to the nearest whole percent or whole number.

1. Tremler has 6,800 inquiries resulting in 2,350 sales. What percent of the inquiries result in sales? ________
2. Y-Sale has 4,130 inquiries resulting in 510 sales. What percent of the inquiries result in sales? ________
3. Siesta has 2,160 inquiries resulting in 600 sales. What percent of the inquiries result in sales? How many inquiries are needed to make 1,500 sales? ________
4. Awesome has 890 inquiries resulting in 215 sales. What percent of the inquiries result in sales? How many inquiries are needed to make 400 sales? ________
5. A company says that 62% of inquiries result in sales. How many inquiries would be needed for 500 sales? for 2,000 sales? ________

Directions: Solve problems 6–10, given the average profit per sale. Round sales up.

6. A convention costs a company a total of $2,561. It earns a $3.94 profit on each sale. How many sales are needed to cover the convention costs? ________
7. Kenton Co. pays a total of $5,052 in convention costs. The profit on each sale is $11.21. How many sales are needed to cover the convention costs? ________
8. Hot Stoves pays $3,480 for a booth and $2,350 for oven mitts to give away. It pays 8 people $145 each to staff the booth. The profit per sale is $395. How many sales are needed to cover convention costs? ________
9. Good Times pays $2,930 for a booth and $3,172 for frames to give away. It pays 12 people $115 each to staff the booth. The profit per sale is $214. How many sales are needed to cover convention costs? ________
10. JP Co. pays $1,350 a day for a booth and gives away 2,400 magnets that cost $.61 each. It pays 9 people $150 each per day. The profit on each sale is $1.92. How many sales are needed to cover the 3-day convention costs? ________

Name Date Period

More Conventions and Exhibits Practice

EXAMPLE P & J expects 90 people an hour to visit its booth. Each visitor takes about 3 minutes of an employee's time. How many employees are needed at the booth each hour?

Step 1 Find the total number of minutes per hour with visitors.

90 people × 3 minutes = 270 minutes

Step 2 Divide the total number of minutes by the minutes in an hour. Round up.

270 minutes ÷ 60 minutes = 4.5 ≈ 5

P & J needs 5 employees at the booth each hour.

Directions: Find the number of employees needed each hour. Round up.

	Visitors per Hour	Time with Each Visitor	Employees Needed per Hour
1.	120	4 minutes	
2.	50	2 minutes	
3.	85	2.5 minutes	
4.	130	3.5 minutes	
5.	210	4.5 minutes	

Directions: Solve problems 6–10.

6. A company gives away 30 basketball tickets. Each ticket costs $35. How much is the entertainment expense? __________

7. A company takes 58 people to dinner for $45.95 per person, including tip. How much is the entertainment expense? __________

8. A company takes 27 people to dinner for $28.76 each. It gives away 25 theater tickets that cost $85 each. How much are the entertainment costs? __________

9. A company gives away 5 vacation trips for $1,280 each and 15 dinner certificates for $150 each. How much are the entertainment costs? __________

10. A company spends $8,467 for a banquet and 30 identical door prizes. The banquet costs $5,692. How much does each door prize cost? __________

Marketing Channels

EXAMPLE Tighten Inc. advertises a wrench in *Nuts & Bolts Magazine* for an advertising cost of \$62,000. *Nuts & Bolts* has a distribution of 810,000 readers. Tighten expects that $\frac{1}{50}$ of the readers will buy the wrench. What is the advertising cost for each wrench sold by the advertisement?

Step 1 Find the number of sales. $\frac{1}{50} = 1 \div 50 = .02$ $810{,}000 \div .02 = 16{,}200$

Step 2 Divide the advertising cost by the number of sales. Round to the nearest cent.
$\$62{,}000 \div 16{,}200 \approx \$3.83$

Advertising in *Nuts & Bolts Magazine* costs \$3.83 for each wrench sold by the advertisement.

Directions: Find the advertising cost per sale. Round to the nearest cent.

	Magazine Distribution	Cost of Advertising	Fraction of Readers Who Buy	Advertising Cost per Sale
1.	500,000	\$28,000	$\frac{1}{100}$	
2.	460,000	\$53,000	$\frac{1}{50}$	
3.	800,000	\$74,000	$\frac{1}{40}$	
4.	760,000	\$130,000	$\frac{3}{50}$	
5.	3,900,000	\$520,000	$\frac{9}{100}$	

Directions: Use the table to solve problems 6–7.

6. A company wants to sell a product to men aged 37–50. Which store has the most potential buyers?

____

7. A company wants to sell to men and women aged 26–36. Which two stores should sell its products?

____

Age	Gender	Store A	Store B	Store C
15–25	F	16%	38%	46%
15–25	M	61%	21%	18%
26–36	F	6%	43%	51%
26–36	M	52%	35%	13%
37–50	F	1%	65%	34%
37–50	M	49%	31%	20%

Directions: A marketing plan is effective if it costs less than the given percent of profits from new sales. Write *Yes* or *No* to tell whether each marketing plan is effective.

	Cost of Marketing Plan	Sales	Profit as a Percent of Sales	Amount of Profit	Percent of Profit from New Sales	Maximum Amount for Effective Advertising	Effective?
8.	\$9,500	\$618,000	30%		10%		
9.	\$134,000	\$2,300,000	29%		25%		
10.	\$218,000	\$1,450,000	16%		47%		

Name Date Period

Chapter 13, Lesson 1

Workbook

56

Catalogs and Customer Service

EXAMPLE Peterson's catalog has 110 pictures that cost $210 each. Each picture has a description that costs $40. Peterson spends $127,200 to design and print the catalog. It spends $145,100 on a mass mailing. Peterson mails 180,000 catalogs to potential customers. What is the cost to produce and mail each catalog?

Step 1 Find the cost of the pictures and the descriptions.
110 pictures × $210 = $23,100 110 descriptions × $40 = $4,400

Step 2 Find the total cost of the catalogs.
$23,100 + $4,400 + $127,200 + $145,100 = $299,800

Step 3 Divide the total cost of the catalogs by the number of catalogs. Round to the nearest cent. $299,800 ÷ 180,000 ≈ $1.67

It costs Peterson $1.67 to produce and mail each catalog.

Directions: Find the cost per catalog. Round to the nearest cent.

	Catalogs Mailed	Number of Pictures and Descriptions	Cost per Picture	Cost per Description	Designing/ Printing Cost	Mailing Cost	Cost per Catalog
1.	300,000	60	$185	$48	$85,915	$92,235	
2.	600,000	80	$220	$50	$348,916	$296,804	
3.	280,000	120	$275	$70	$128,230	$152,641	
4.	830,000	250	$265	$54	$538,919	$685,234	
5.	990,000	320	$165	$65	$625,913	$715,504	

Directions: Plan A charges $15 per month per phone plus $.08 per minute, and charges for fractions of a minute. Plan B charges $8 per month per phone plus $.13 per minute, and a fraction of a minute counts as a full minute. Find each cost for one month. Round to the nearest cent.

	Phones	Calls	Average Call	Plan A Cost	Plan B Cost
6.	18	250	1 minute, 30 seconds		
7.	32	600	3 minutes, 30 seconds		
8.	45	587	2 minutes, 15 seconds		
9.	28	1,951	3 minutes, 5 seconds		

Directions: Solve problem 10.

10. A business orders 23 computers for $2,028.35 each and a server for $107,492.86. What is the total cost? ______________

Name Date Period

Shipping Orders

EXAMPLE Sharon orders 16 beakers for $6.85 each from a catalog. The shipping charge is the greater of $5\frac{1}{2}$% of the total cost or $8.50. How much does Sharon pay for shipping?

Step 1 Find the total purchases. $16 \times \$6.85 = \$109.60$

Step 2 Find the percentage of the purchases. Round to the nearest cent.
$\$109.60 \times 5\frac{1}{2}\% = \$109.60 \times 5.5\% = \$109.60 \times .055 = \$6.028 \approx \$6.03$

Step 3 Choose the greater shipping charge. $\$6.03 < \$8.50$
Sharon pays $8.50 for shipping.

Directions: Solve problems 1–3. Round to the nearest cent.

1. Melvin orders dishes that cost $815.89 from a catalog. The shipping charge is the greater of 8% of the total cost or $35. How much does Melvin pay for shipping? ____________
2. Sam orders a couch that costs $658 from a catalog. The shipping charge is the greater of $3\frac{1}{2}$% of the total cost or $50. How much does Sam pay for shipping? ____________
3. Tina orders 15 gifts that cost $38.95 each from a catalog. The shipping charge is the greater of $6\frac{1}{2}$% of the total cost or $25. How much does Tina pay for shipping? ____________

The table shows rates for 2-day delivery from Cheyenne, Wyoming, to Portland, Oregon.

Carrier	Up to 1 Pound	More Than 1 Pound to 2 Pounds	More Than 2 Pounds to 3 Pounds	More Than 3 Pounds to 4 Pounds
A	$8.20	$9.21	$10.43	$11.75
B	$9.10	$10.14	$11.44	$13.26

Directions: Use the table above to solve problem 4.

4. A shipment has 17 packages weighing 3 pounds, 1 ounce each. Which carrier costs less? how much less? ____________

Costs to Ship 6 Pounds Between Zip Codes 35630 and 66101			
Guaranteed Delivery Time	**Next-Day Service**	**Two-Day Service**	**Three-Day Service**
8:30 A.M.	$52.93		
12:00 P.M.	$31.39	$ 13.57	
End of Day	$27.34	$ 12.05	$6.05

Directions: Use the chart above to solve problem 5.

5. Carla decides to use 2-day service by 12:00 P.M. How much does shipping a 6-pound order cost? ____________

Name Date Period

Handling and Processing Orders

EXAMPLE A company packs an order in a box 12 inches long by 12 inches wide by 12 inches high. Wrapping the order uses 2.4 square feet of packing material at $.45 per square foot. Handling the order takes an employee 30 minutes. The employee costs $19.50 per hour, including benefits. What is the total cost to process and handle the order?

The table shows the company's cost for packages of 10 boxes.

Length (inches)	Width (inches)	Height (inches)	Cost for 10 Boxes
12	12	12	$8.80
18	12	16	$9.60
20	20	20	$12.40

Step 1 Find each cost. Box: $\$8.80 \div 10 = \$.88$ Packing Material: $\$.45 \times 2.4 = \$1.08$

Employee: $\$19.50 \times 30 \text{ minutes} = \$19.50 \times \frac{30}{60} \text{ hour} = \$19.50 \times .5 \text{ hour} = \$9.75$

Step 2 Add the costs. $\$.88 + \$1.08 + \$9.75 = \$11.71$

The order costs $11.71 to process and handle.

Directions: Use the table above to find the total cost of processing and handling. Round to the nearest cent.

	Box Size (inches)			Packing Material ($.47 per square foot)	Labor Time ($16.80 per hour)	Total Cost of Processing and Handling
	L	W	H			
1.	20	20	20	1.5 square feet	30 minutes	
2.	20	20	20	2.5 square feet	45 minutes	
3.	12	12	12	1 square foot	30 minutes	
4.	12	12	12	1.5 square feet	15 minutes	
5.	18	12	16	2 square feet	10 minutes	

Directions: Anne wants to order 2 yellow pairs of pants in small from page 51 of a catalog. The item number is BOD024. Each pair of pants costs $48.92. Complete the order form. Round to the nearest cent.

Catalog No.	Quantity	Page No.	Description	Size S, M, L	Color	Unit Price	Total Price
6.							

Subtotal	7.
Sales Tax on the Subtotal—Add 8%	8.
Shipping/Handling Charges—Add 7%	9.
Total	10.

Inventory and Warehousing

EXAMPLE About $\frac{2}{3}$ of a store's catalog sales are canceled when the items are back ordered. In August the back-ordered items are worth $27,102.36 in sales. How much money in sales will be lost if $\frac{2}{3}$ of the orders are canceled?

Multiply to find the loss in sales.

$$\frac{2}{3} \times \frac{27{,}102.36}{1} = \frac{2}{\cancel{3}_1} \times \frac{\cancel{27{,}102.36}^{9{,}034.12}}{1} = 2 \times 9{,}034.12 = 18{,}068.24$$

The store loses $18,068.24 of $27,102.36 in back-ordered sales.

Directions: Solve problems 1–10.

1. S & N has $19,180.24 worth of sales back ordered. How much money in sales will be lost if $\frac{1}{4}$ of the orders are canceled? ________
2. New Times has $21,475.60 worth of sales back ordered. How much money in sales will be lost if $\frac{2}{5}$ of the orders are canceled? ________
3. Krunch has $32,614.80 worth of sales back ordered. How much money in sales will be lost if $\frac{3}{8}$ of the orders are canceled? ________
4. Renton Reels has $8,597.31 worth of sales back ordered. How much money in sales will be lost if $\frac{1}{3}$ of the orders are canceled? ________
5. Pots and Pans has $48,356.50 worth of sales back ordered. How much money in sales will be lost if $\frac{4}{5}$ of the orders are canceled? ________
6. Try This has $30,480.60 worth of sales back ordered. How much money in sales will be lost if $\frac{2}{3}$ of the orders are canceled? ________
7. Delightful Things has $58,600.80 worth of sales back ordered. How much money in sales will be lost if $\frac{5}{8}$ of the orders are canceled? ________
8. Saturday Excursions has $46,925.64 worth of sales back ordered. How much money in sales will be lost if $\frac{3}{4}$ of the orders are canceled? ________
9. B & T lost $8,300.67 in sales because $9,486.48 worth of sales was back ordered. What fraction of sales was canceled? ________
10. Nice Things lost $5,606.28 in sales because $12,458.40 worth of sales was back ordered. What fraction of sales was canceled? ________

Name Date Period

More Inventory and Warehousing Practice

EXAMPLE A mail-order business sells home decorations. Storage costs $.006 per cubic foot per day. A statue that measures 3 feet long by 2 feet wide by 2 feet, 4 inches high is in inventory. How much does storing the statue in inventory for 30 days cost?

Step 1 Find the volume of the storage space. Change all measurements to feet.

2 feet, 4 inches = $2\frac{4}{12}$ feet = $2\frac{1}{3}$ feet

Use the formula $V = lwh$. $l = 3$ feet $w = 2$ feet $h = 2\frac{1}{3}$ feet

$V = 3 \times 2 \times 2\frac{1}{3} = \frac{3}{1} \times \frac{2}{1} \times \frac{7}{3} = \frac{42}{3} = 14$

Step 2 Find the storage cost for one day. Multiply it by 30.

14 × $.006 = $.084 $.084 × 30 = $2.52

It costs $2.52 to store the statue for 30 days.

Directions: For each cost per day, find the volume and cost of storage. Use the formula $V = lwh$. Round the volume up to the next whole cubic foot.

	Cost per Cubic Foot per Day	Length	Width	Height	Volume	Cost for 30 Days
1.	$.003	3 feet	5 feet	2 feet		
2.	$.007	2 feet	2 feet	2 feet		
3.	$.012	4 feet	4 feet	5 feet		
4.	$.008	3 feet	3 feet	5 feet		
5.	$.023	2 feet, 6 inches	3 feet, 6 inches	2 feet		
6.	$.005	3 feet, 2 inches	4 feet, 8 inches	2 feet, 6 inches		

Directions: Use a Monday–Saturday calendar to solve problems 7–10.

7. A warehouse manager takes 2 days to reorder items. The supplier needs 3 days to fill the order. Shipping takes 3 more days. The manager starts reordering on Monday. When will the order arrive? ______________

8. A warehouse manager takes 3 days to reorder. The supplier needs 3 days to fill the order. Shipping takes 4 more days. The manager starts reordering on Tuesday. When will the order arrive? ______________

9. A warehouse manager takes 2 days to reorder. The supplier needs 3 days to fill the order. Shipping takes 2 more days. The order arrives on Wednesday. When did she place the order? ______________

10. A warehouse manager takes 3 days to reorder. The supplier needs 3 days to fill the order. Shipping takes 3 more days. The order arrives on Saturday. When did he place the order? ______________